A Guide to...

Incubation & Handraising Parrots

by Phil Digney

Published and Edited by ABK Publications ©

First Published 1998 by
ABK Publications
PO Box 6288,
South Tweed Heads,
NSW. 2486. Australia.

ISBN 0 9587102 1 X

Front Cover:
Top left: Newly fledged Major Mitchell's Cockatoo chick.
Centre: Handfed White-tailed and Major Mitchell's Cockatoo chicks.
Bottom left: Newly hatched Major Mitchell's Cockatoo.
Bottom right: Syringe feeding an Alexandrine Parrot chick.
Back Cover: Cockatiel chick.

All cover photographs by Phil Digney

ABOUT THE AUTHOR

Phil Digney has had a lifelong passion for parrots, with a particular interest in the endangered species. Born in 1967, he grew up in the heart of the jarrah forests of south-west Western Australia, in the small town of Collie. With most of his spare time (and some of his school time!) spent along the Collie River and in the surrounding bushland, he soon became fascinated by the local parrots. By the age of ten he was handraising Western Rosellas and Red-capped Parrots and since then has kept and successfully bred most Australian species of parrots as a private aviculturist.

At the age of 21 after many months of research and compiling of old records, he embarked on a six month field search for the supposedly extinct Paradise Parrot, centered in mid-Queensland. Upon completion of this search he took up the position of bird keeper in the incubation/handraising complex of Pearl Coast Zoo, Broome. At that time (1990), Pearl Coast Zoo maintained the largest and most diverse collection of parrots and cockatoos in Australia and it was there that Phil's incubation and handraising skills developed on a professional level, not only with Australian species but also with many foreign species.

Upon the closure of Pearl Coast Zoo in 1991 he immediately took up the position of curator at Rainbow Jungle, Western Australia's largest parrot park and breeding centre. After two years as curator he then devoted another four months in the field to following up new and promising reports of the Paradise Parrot. While neither search uncovered a Paradise Parrot, Phil remains convinced that a few of what has been described as 'the world's most beautiful parrakeet', still survive.

After the second search, Phil returned to Rainbow Jungle as curator for a further two years. Incubation and handraising were central to the breeding success of Rainbow Jungle, particularly with the cockatoos and the four years there provided Phil with the opportunity to gain experience with various incubators, handraising formulas, feed methods etc. During 1996, owner of Rainbow Jungle, Ian Brumley and Phil initiated a three year White-tailed Black Cockatoo (Carnaby's) program with the Department of Conservation and Land Management (CALM). The program objective was to establish a viable captive population of this threatened species, thereby securing their future and at the same time reducing pressure on the wild stock from trappers and nest raiders. Over the three year period a considerable number of eggs and chicks were removed from the wild and under CALM supervision, distributed to five private aviculturists, including Rainbow Jungle and Phil.

At the end of 1996, Phil moved back to Perth where he continues to maintain a private collection of cockatoos while providing a variety of services to avicultural circles including incubation/handraising, deworming, aviary building etc. Phil is also an experienced bird taxidermist.

Phil continues to be a regular writer/photographer for ***Australian Birdkeeper Magazine*** and also authors articles for various smaller parrot journals around Australia. He spoke at the Australian, AFA Avicultural Convention in Perth '95 on the theme *Professional Handraising* and was keynote speaker at the South Australian mini convention February '98, speaking on the topics *Handraising* and *White-tailed Black Cockatoo Program - securing their future.*

Phil believes that aviculturists will play an increasingly important role in the preservation of many bird species that are currently under threat in the wild and whenever possible promotes aviculture as an invaluable conservation tool. He continues to spend as much time as possible photographing and enjoying parrots at their best - in the wild. He is constantly looking for new challenges and experiences with his favourite creatures - parrots.

ACKNOWLEDGEMENTS

The creation and development of this title has been a very challenging and immensely satisfying experience on a personal level. However, many people have assisted me in completing this project. Whether they allowed me into their aviaries to take a particular photograph, provided me with their weight charts, offered their thoughts and experience on certain topics or one of the many other tasks too numerous to mention, I would sincerely like to thank the following:

Diana Andersen
Dave Bourke
Bob Branston
Dr Danny Brown
Dr Ray Butler, Risely Veterinary Centre
Stuart Chamberlain, Bird & Fish Place
Geoff Coombes
Shane Drew and Ryan Watson, Birdworld
Kevin Evans and Chris Hibbard, Taronga Zoo, Sydney
Kevin Gobby
Neil Hamilton, Perth Zoo
George Hobday
Hank Jonker
Russell Kingston
Chris Martin
Tony Nelson
Dorothy Payne
Bevan Reynolds
Cheryl Thode
Stewart Williamson

Particularly, I would like to thank Ian and Lisa Brumley of Rainbow Jungle, for it was there that I was given the freedom and opportunity to develop much of the experience I now have. Finally, my thanks to Nigel Steele-Boyce and Sheryll Stevens of ***ABK Publications*** for inviting me to author this book in the first place and having faith in me.

Phil Digney.

Incubation

Handraising Parrots

INTRODUCTION

Australia has long been styled 'the land of parrots', but for how much longer? Many of these unique creatures that have provided us with so much pleasure for so long are now in danger of disappearing. The world is changing and with these changes have come the swallowing up of much of their natural habitat.

Two Australian species are missing, presumed extinct, while many others are declining, the Glossy Black Cockatoo, the Swift Parrot, the Major Mitchell's Cockatoo, the Ground Parrot ...

Fortunately, aviculture is changing also. Gone are the days where the breeder provides a pair of parrots with a nesting log and hopes that a bowl of dry seed and a large slice of luck will produce chicks. In fact, such have been the advances in knowledge, management and breeding success of Australia's parrots in captivity that aviculture is being applauded as more than just a hobby, rather a skill. The employment of aviculturists by Government conservation bodies in their efforts to preserve species such as the White-tailed Black Cockatoo, Naretha Blue-bonnet and the Orange-bellied Parrot, are testimony to this ability.

The hallmark of the true aviculturist is successful breeding and we now have at our disposal all the tools needed to make the captive breeding of any species a reality. As aviculturists, it becomes our duty to gather as much knowledge and experience as possible so that in some way, each of us may have something to offer these vulnerable creatures.

Artificial incubation and handraising have been the cornerstones of this successful breeding in modern aviculture and while the reasons for these two practices are as varied as the species we keep, surely the primary one is improved breeding results. In dealing with these two complimentary subjects the objectives of this book are twofold.

Firstly, to equip the breeder with the necessary knowledge to enjoy the immense satisfaction that comes from successful incubation and handraising of parrots and secondly, it is to give something back to the parrots.

Incubation

ARTIFICIAL INCUBATION

The reasons for artificial incubation fall into two categories - necessity and choice. Some hens lay their eggs on the ground despite the presence of a suitable nest. Other hens abandon their clutch halfway through incubation for no apparent reason. Occasionally, if the species is one in which both sexes incubate - for example, Major Mitchell's or Gang Gang Cockatoos - one of the birds may be an egg breaker. Alternatively, there are those aviculturists who manage expensive or rare species and make the choice of deliberately pulling a clutch of eggs to induce the hen to re-lay.

Above: Successful incubation begins with a quality diet including fruit and vegetables. The nutrition in the egg yolk on which the embryo feeds will be directly determined by the nutrition fed to the breeders. Keep it fresh and varied.

Below: Eclectus require 70-80% of their diet to be fruit, vegetables and nuts to breed successfully.

Whatever the reasons, artificial incubation picks up where nature left off. It is a challenging, exciting and fulfilling pastime. There will always be something special and satisfying in watching a chick poke that first hole in the shell and emerge from the egg that was artificially incubated from day one.

The technology applied to the incubators and hatchers available today, combined with the documented findings of experienced and know-ledgeable aviculturists, has taken much of the anxiety out of artificial incubation and in many senses it is now a simple procedure. However, like most things in life there is a learning curve and raw information is no substitute for experience. Reading books such as this and absorbing as much information as possible before embarking, is an excellent start, though to be truly successful, the practical application of that knowledge is irreplaceable. Sure, there will be problems and there will be frustrating moments as one ponders an unexplained death, but through the keeping of accurate records, careful monitoring of your instruments and working strictly within established parameters, problems will be rare and losses even rarer. Learn from your mistakes, finetune your techniques, and it won't be long before you *can* count your chicks before they hatch!

MEET THE EGG!

To become efficient in the art of incubation it is essential to understand the basic components and functions of the central figure - the egg. The egg is a tiny nursery within which a series of amazing changes take place as a few cells develop into a living, breathing, fully formed chick. By understanding the five major components of the egg, many of the principles of artificial incubation are better appreciated.

The Shell

The shell consists of two layers; the inner mammillary layer and the outer 'spongy' layer. Despite the appearance of being a sealed unit, under magnification these two shell layers are found to be porous. These pores allow the developing egg to transpire water and gases over the incubation period which are necessary functions if the chick is to hatch successfully and unaided. Coating the entire surface of the shell is a thin film or 'bloom' which regulates the evaporation rate of the egg's moisture.

There is a very high calcium content in the egg shell and laying of several eggs significantly depletes the hen's calcium reserves. With this in mind, calcium supplementation is important in preventing eggbinding and soft/thin shelled eggs. Supplementation should begin prior to the breeding season in the form of liquid in water eg Calcivet™ or similar product, powder sprinkled on greens or a regular supply of cuttlefish and fine grit. There are a number of proprietary brands of calcium supplements available in liquid, powder or tablet form.

The Shell Membranes

There are two shell membranes; the inner and the outer. They lie alongside each other directly beneath the surface of the shell. The inner membrane is separated from the outer at the blunt end of the egg, forming an air-cell. In the days following lay, whether developing or not, the air-cell slowly expands as moisture escapes the shell and draws the inner membrane further down. If an egg is found unexpectedly in the nest, the size of the air-cell will provide a good indication of its age. In an old egg, the air-cell will occupy up to 50% of the space within the egg whereas in some fresh eggs the air-cell is barely detectable.

The Albumen

The albumen or eggwhite is a clear fluid found inside the freshly opened egg. It consists of an outer layer of thin white, a middle layer of thick white and an inner thinner layer which surrounds the yolk itself. Both ends of the yolk are anchored to the shell membrane within this albumen by a twined structure called chalazae, which gives the egg stability and absorbs sudden movements.

The Yolk

The yolk is the food of the developing embryo. Lighter than the albumen, it floats to the top of the egg, and contains all the nutritional elements necessary to produce a healthy chick. For this reason it is essential to provide the laying hen with a varied, high quality diet that can be passed on to the chick via the yolk.

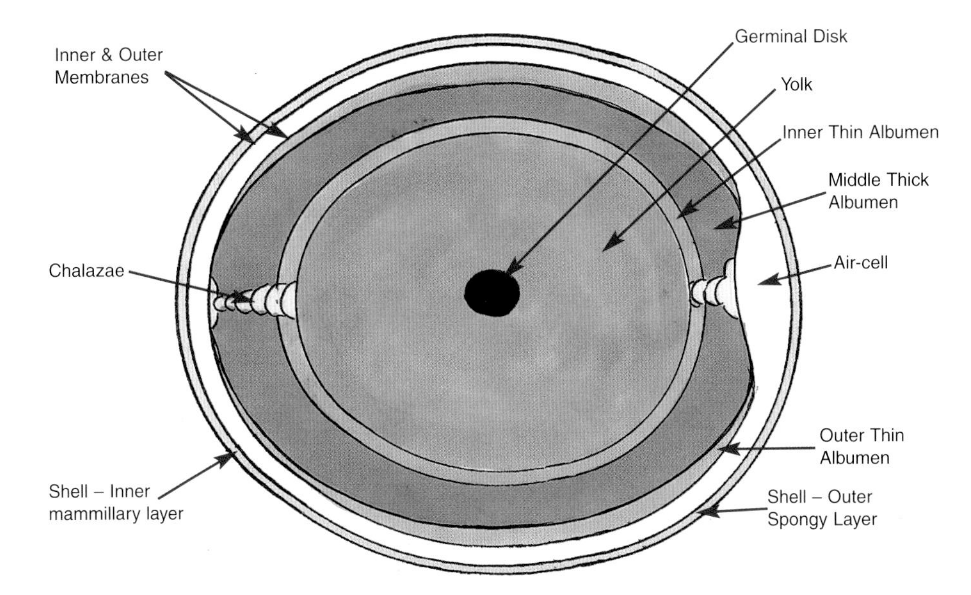

Left: Growing as many greens as possible yourself is the best way to go. Whatever the source of the greens, wash thoroughly before feeding.

The Germinal Disc

The germinal disc appears as a small white dot on the upper surface of the yolk. If fertilisation has occurred and conditions are right, this dot or blastodisc begins to develop into an embryo. Consistent turning of the egg prevents this developing embryo from sticking to the egg membrane.

The egg, then, is deceptively complex. What seems to be a very simple structure, is in fact a sophisticated storehouse of life, providing the embryo with all the protection and resources it will need before finally emerging as a healthy chick. Hopefully this progression will be seen in the majority of eggs placed in the incubator - and why shouldn't it be? Provided the birds are not too old or related, are fed a quality diet and have plenty of room in which to exercise, infertile eggs will be rare.

CHOOSING AN INCUBATOR/HATCHER

Incubators

There are almost as many models of incubator available today as there are species of parrot. The price range starts from under AUD$500 to over AUD$2000. The varying price tags, however, should not blind you to the fact that basically, they all perform the same function - supplying regulated heat to the egg in an insulated environment. Research as many as you can and talk to experienced breeders in regard to the particular models they use. When purchasing the incubator there are two basic options available:

- Hand-turn versus Auto-turn
- Fan Forced (moving air) versus Still Air

Hand-turn versus Auto-turn

While a unit that turns the eggs automatically commands a somewhat higher price, it is worth the extra money. Parrot eggs need to be turned at least three times a day for almost the entire duration of the incubation process, so unless you have plenty of spare time and are constantly in attendance, an auto-turn model is the wiser choice. Manual turning soon becomes tedious and time consuming, providing no realisable benefits. While it has been claimed that the auto-turn model does not provide a natural movement (in essence true) there is no evidence to suggest that it produces a lesser quality of chick, or a higher death rate. Some of the cheaper units hold the egg in a rack that can be manually turned from outside the unit, and although this saves individual turning of the eggs (and thereby handwashing), it still requires that you attend the incubator several times a day.

Right: A popular fan forced, auto-turn incubator, the Brinsea™ Octagon 20 MK III.

Auto-turn models are the preferred choice of an increasing number of aviculturists, therefore as the demand increases so does the quality of the unit. However, two factors still need to be considered when contemplating an auto-turn model. The first is that vibration from the turning mechanism may damage the small embryo. Indeed, this may be true of older models, however, modern units are virtually vibration free. The second factor is the quality of the turning or rolling action. Jerking or sudden rocking of the eggs can be just as detrimental as vibration, therefore, thoroughly assess the unit for its performance before purchase. All models will need monitoring once a day to ensure that humidity, temperature and development are satisfactory, however, the auto-turn system will greatly reduce your workload.

Above: Another popular incubator, the Marsh™ Rolex, fan forced, auto-turn unit.

Below: An excellent still air hatcher, the Hatchmaker by Brinsea™.

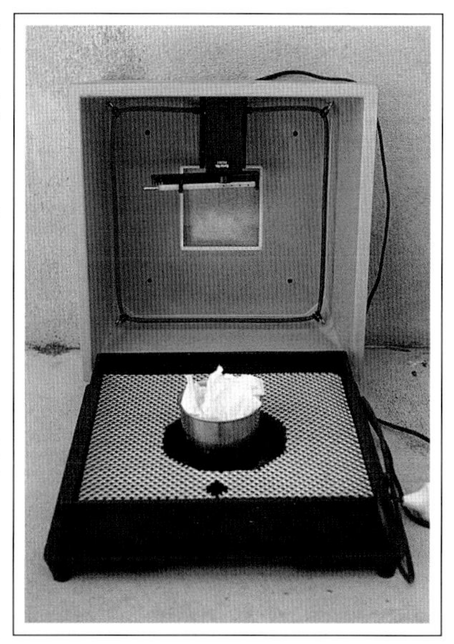

Fan Forced versus Still Air

A fan forced (moving air) model will alleviate a problem often encountered in still air models, variable temperature zones. Still air units may not only vary in temperature at different levels, they can also contain zones which depending on their proximity to the vent holes, differ considerably in temperature from the thermometer location. Variations of up to 3°C have been recorded in a few still air units, and this is really not suitable for artificial incubation of the parrot egg. A fan forced model will distribute the warm air evenly throughout the unit, providing a much more stable environment.

So, when choosing an incubator, consider whether you are prepared to spend a little more money, to avoid potential problems along with extra work. Successful incubation relies on accurate temperature control and smooth regular turning and while there is no such thing as 'set and forget', it can be very helpful to have a reliable incubator to do the work for you. Two of the many models that meet the suggested criteria and are finding favour in Australian aviculture are the Brinsea™ Octagon 20 Mark III and the Marsh™ Rolex. Their middle-of-the-range price tag, smooth turning motion and precise, fan forced temperature control make them ideal units for all levels of incubation. Another brand of incubator, that although more expensive, proving to be popular and reliable is the AB™ Newlife 75 MK4 Incubator.

Hatchers

The hatching process generally begins approximately five days before the expected hatch date, with what is termed the *drawdown* (discussed under *Hatching Details*). Internal pip then occurs three to four days prior to expected hatch date. It is at this point that the egg no longer requires turning and now needs a higher level of humidity. If you have more than one egg, you will require a second unit to be available for the hatching egg.

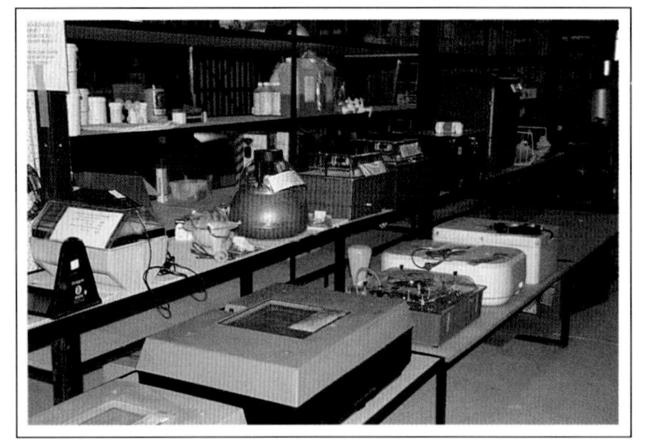

Left: There are now a large range of incubators to choose from.

It is true that incubators *can* double as hatchers, by the raising of the humidity and the placement of the egg somewhere within the unit where it will not be turned, but this presents problems. The higher humidity will be detrimental to less-developed eggs and some auto-turn units have nowhere to place an egg that does not need turning. Some aviculturists who have not had a separate hatcher on hand, have had to leave a hatching egg in normal incubation humidity, and while chicks *have* successfully hatched this way, it is far from ideal and will be discussed further in the section *Hatching Details*. At the end of the day, a separate unit specifically for hatching is far more preferable.

Consider the fact that regular inspection is necessary during hatching. Look for a unit that provides clear viewing of the egg through the cover, ideally one in which the cover does not have to be fully removed when removing the egg for candling. A hinged lid or sliding front cover retains more heat within the unit than a cover that is lifted entirely to one side. Depending on how frequently the egg is candled and examined, this continual loss of warm air may weaken the chick as it works on the shell.

Hatchers are available as fan forced (moving air) or still air units. Some aviculturists prefer to hatch in a still air environment, believing that fan forced air may dry out the chick and membranes as hatching takes place, causing the chick to become stuck inside the shell. However, it is not really possible for this to happen, given the high level of humidity that the hatcher provides. The only real problem with fan forced units lies with the operator, not the machine. In fan forced hatchers the water jackets dry out at a faster rate than still air units, therefore, if the aviculturist allows the unit to run dry, then yes, the chick may dry out and become caught. Regular monitoring of the water supply prevents this problem, while the fan maintains a uniform air temperature. Two popular fan forced hatchers are the AB™ Startlife 25 Hatcher or the Brinsea™ Octagon 20 Parrot Rearing Module.

This is not to say that still air hatchers are inferior - in fact, many chicks are successfully hatched every year in such units. An excellent still air unit designed specifically for hatching is the Brinsea™ Hatchmaker. It has a clear viewing plate in the lid and by hinging the lid at the

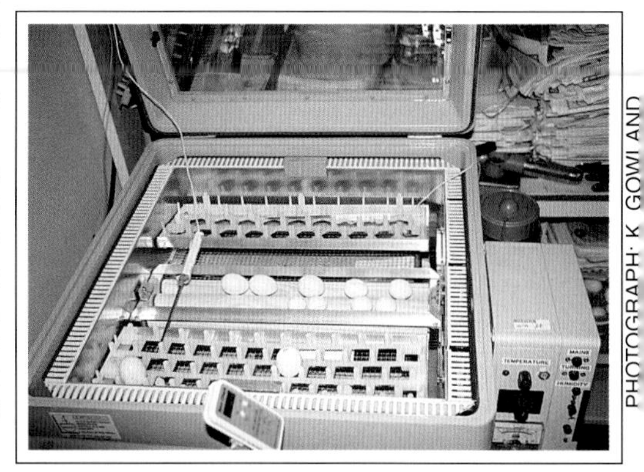
PHOTOGRAPH: K. GOWLAND

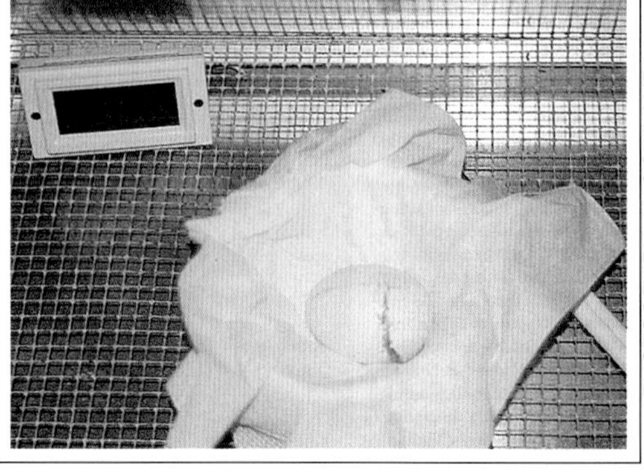
PHOTOGRAPH: K. GOWLAND

Top right: An AB™ Newlife 75 MK4 Incubator.
Bottom right: A chick hatching out of the egg is the ultimate sign of successful artificial incubation.

PHOTOGRAPH: K. GOWLAND

Left: AB™ Startlife 25 Hatcher

rear, the eggs can be removed for candling without significant loss of temperature and humidity. This unit has largely overcome the problem of variable temperature zones by supplying heat from all four sides of the unit, and halfway down. The thermometer can be located to sit directly alongside the hatching eggs, to ensure an accurate reading at the egg location.

The purchase of a suitable incubator and hatcher can be a confusing time for the inexperienced. It is advisable to talk with other aviculturists and equipment suppliers to find out the advantages and disadvantages of various models. The final decision will depend on your budget, other aviculturists' opinions, your lifestyle and the number of eggs to be incubated and hatched. Take your time, do a little research, and the result will be an efficient, reliable unit that will suit your needs and produce many a healthy chick.

An important note here - always have your incubator and hatcher units tested and running efficiently two to three weeks prior to the breeding season to ensure that everything is working well and to allow you to finetune the settings prior to placement of eggs.

EQUIPMENT

Having purchased an incubator and hatcher, you must now provide yourself with some basic items that should be on hand before incubation commences. It is no good waiting until a problem arises before buying the equipment that may have prevented it.

Wet Bulb Thermometer/Humidity Gauge

The humidity of the incubator is a crucial aspect of successful incubation and hatching, and an instrument is required to allow the accurate recording of humidity levels. Many incubators come equipped with a humidity dial or gauge of some description - some quite accurate, others hopelessly erroneous. Traditionally the most reliable instrument has been the wet bulb thermometer. This device consists of a thermometer with a cloth wick attached to the bulb end, which is hung in a vial of water. It is basic in design, easy to read and extremely accurate. More recently, various digital and mechanical humidity readers have become available, but before relying on these, one must calibrate them against a wet bulb thermometer.

One advantage of the more accurate digital readers is the fact that they also record the minimum and maximum temperatures reached between personal visits to the hatcher. The beauty of this is that you will be provided with any fluctuations occurring in your absence, so unacceptable variations can be noted. Should the wet bulb thermometer be the instrument of your choice, be aware that the water in the glass vial will need regular topping up. Left to run dry, the

Wet bulb thermometer for humidity readings. Simple and very accurate.

thermometer will shoot up to the dry temperature and scare the daylights out of you next time a reading is taken. The wick should be replaced every few weeks, however a regular good clean with soapy water to maintain accuracy, is recommended.

NOTE: A wet bulb thermometer only records the humidity temperature, not the dry temperature.

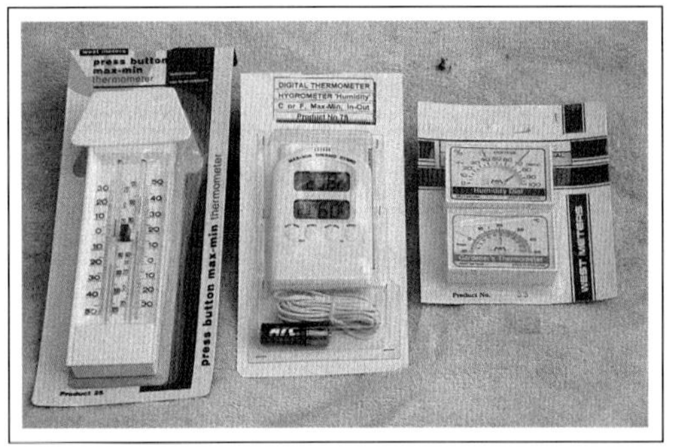

Above: There are now a variety of humidity readers available. The digital models have the advantage of recording high and low variations in your absence.
Below: Candlers, essential for egg examination, come in all shapes and sizes. The black one is a 240 volt, plug in model. The smaller one is battery operated with a flexible neck.
Bottom right: Before incubation begins a variety of tweezers and scissors need to be on hand.

Candler

A candler is simply a purpose-built torch or light with a concentrated beam for examining the internal developments of the egg. This device is essential not only for determining fertility during the first few days of incubation, but also for regular examination throughout the process. Candlers come in all shapes and sizes, from the small pencil-torch models to the 240 volt hand held lamp. Many parrot dealerships and poultry equipment suppliers carry a range to suit all needs.

Tweezers

It pays to have a couple of pairs on hand. During assisted hatches, tweezers facilitate the breaking away of the shell. Depending on where the chick is positioned in relation to the site being worked on, a blunt pair may be the wisest choice. If a small chick is accidentally stabbed it may bleed to death.

Sterile Water

This is available in small plastic or glass vials, from most chemists. It is used to moisten the inner membrane during assisted hatches, to reveal active blood vessels.

One Millilitre Syringes

Sterile syringes are very cheap and are used to drop sterile water onto the membrane during an assisted hatch.

Heat Source

An external heat source becomes necessary while the egg is outside the hatcher for any length of time during an assisted hatch. The last thing a troubled chick needs is to become chilled as it attempts to emerge from the egg. A desk lamp with a 60-80 watt pearl or coloured globe will provide sufficient warmth and diffused light.

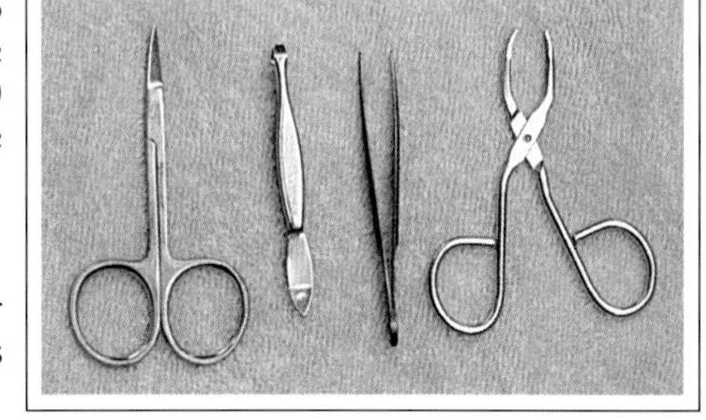

Sterilising Agent

Brooders and hatchers will need regular disinfecting. A recommended agent for this

Above: Avisafe™, a safe and effective disinfectant used for cleaning equipment. Below: Essentials for the treatment of the navel area at hatch time, Betadine™, sterile water, specimen container, cotton wool buds and paint brush.

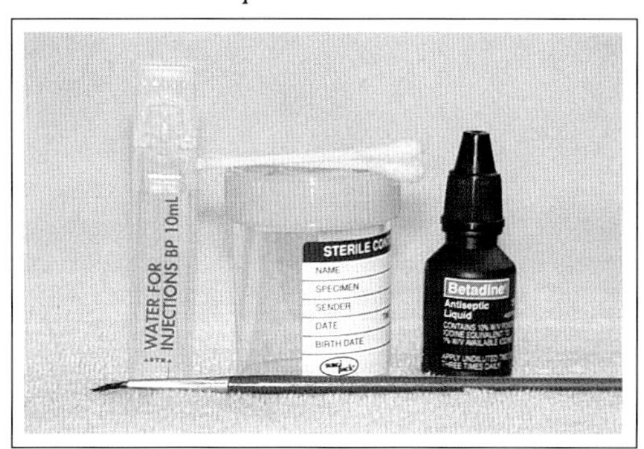

purpose is Avisafe™ by Vetafarm, which is far safer than traditional bleaches.

Distilled Water

Recommended for the water jackets of both the incubator and hatcher.

Betadine™

Available from chemists in 15ml bottles. Diluted 1:10 with sterile water, it is used to treat the navel area of the newborn chick. The mixture is best applied using cotton wool buds or a small paintbrush.

Specimen Containers

Sterile until opened and used for holding sterile water or preparing a mix of Betadine™.

Sterile Gauze

Used in the repair of damaged eggs and available at the chemist.

Cauterising Agent

Silver nitrate sticks or ferric chloride solution applied to cuts etc. to stop bleeding.

Glue

Water-based and non-toxic. Either clear gum glue or PVA woodworking glue for egg repair. Some aviculturists have successfully used clear nail varnish to repair eggs.

Record Book

An essential part of incubation is the keeping of records. Accurate notes on temperature, humidity, changes in the egg and turning times (if hand-turning), will help you to improve your technique. Marginal practices during incubation and over a sustained period will manifest themselves in problem hatches and records will enable you to source the problem. Following is a simple layout that allows you to plot progress.

Right: Cauterising agents, silver nitrate sticks and ferric chloride, are used to stop bleeding in the chick.

EXAMPLE OF TYPICAL INCUBATION RECORDS LAYOUT

Day	Dry Temp. °C	Wet Bulb Humidity °F	Suggested Notes
1	37.2°	82°-83°	Egg laid 27 April (pm). Placed in incubator 28 April (am)
2	37.2°	84°	Humid day
3	37.3°	83°	Hand-turning - 7.00am, 11.00am, 3.00pm, 10.00pm
4	37.2°	83°	etc.
5	37.1°	84°	etc.
6	37.2°	81°-82°	Egg candled - blood vessels present in yolk. Fertile
7	37.1°	82°	etc.
8	37.2°	83°	etc.
9	37.2°	79°	Water ran dry in jackets. Egg beginning to darken.

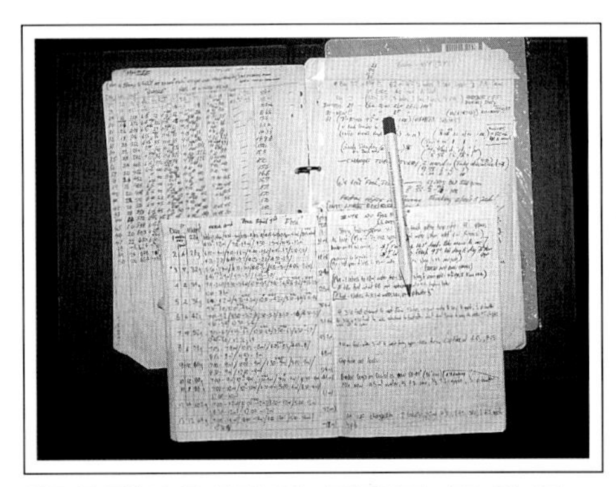

Records become even more critical once hatching begins, therefore it is only by noting the time of each development during the hatching process as it occurs, that you will be able to establish whether a chick is likely to hatch unaided.

Left: Records are an essential part of incubation and handraising.

HANDLING THE EGG

Hygiene

Whether pulling an egg from the nest, transporting from one location to another or manually turning in the incubator, the rule is the same - wash your hands thoroughly with soap and water, or preferably, a powerful cleansing agent such as Hibitane™. Eggs breathe through the pores of the shell and as they cool harmful agents may enter through these pores, particularly if a crack or chip is present. The warm, humid environment of the incubator is the perfect setting for harmful organisms to flourish, so the risk of introduction must be kept to a minimum.

Pulling the Egg from the Nest

It is vital to handle the egg with care when removing it from the nest. An egg damaged or jarred in any way may die during incubation and hours of frustration may be endured agonising over your technique when, in fact, the cause of death was external to the incubator.

The only movement an egg will experience in the nest environment is a gentle roll as the hen turns the eggs with the side of her beak. Any jarring, bumping or

Left: Always wash hands thoroughly with soap or disenfectant before handling eggs.

Above left: Cushion eggs in a bowl of seed whenever transporting.
Above right: When pulling eggs or small chicks from the nest, be aware that some species eg. Eclectus, may dive back into the nest and cause damage.

spinning is unnatural and therefore potentially dangerous to the contents. Do not attempt to pull the eggs while the hen is still on the nest - a torch in the entrance or removal of an inspection hatch will frighten her and the eggs may be damaged. Even if the hen is off the nest, it is important to take care, as some species such as Eclectus and conures will dive back into the nest the moment it is approached or interfered with. Pulling the eggs from the nest while it is hanging is preferred to lowering the log to the ground. Many an aviculturist has slipped or tilted the nest while bringing it down, resulting in broken eggs. An efficient inspection hatch at nest level is an invaluable and recommended addition to the nest.

Before transporting the eggs from the aviary to the incubator room, place them in a bowl of small seed, such as budgerigar mix - and don't trip over the cat on the way! The transportation of eggs by vehicle over long distances has often proved disastrous and should be avoided if at all possible. However, should it become necessary to do so, an excellent and relatively inexpensive 24/12 volt incubator which plugs into the cigarette lighter is the Eco-stat™ Parrot Brooder. Again, the eggs should be cushioned on a layer of seed.

Candling

Candling, or examining the egg contents using a light source, is an essential aspect of the incubation process, as it determines fertility and enables the aviculturist to monitor the development of the chick. Candling is best performed with the egg resting on a small towel, placed on a bench, directly beneath a suitable heat source if the egg is to be outside the brooder for more than a few minutes. Alternatively, the egg can simply be held between the fingers, although care should be taken, as it is surprising how easily it can slip out.

Depending on the type of incubator being used, much of the warm air and humidity will be lost during the recommended daily candling of eggs. Within reason this is acceptable and even desirable as it simulates the natural behaviour of the hen coming off the eggs once or twice a day to feed etc. Eggs in the wild have been unbrooded for up to an hour, gone cold to the touch and yet

Right: Inspection doors, an essential aspect of successful breeding which makes egg removal safer.

successfully hatched. However, this extended period is not recommended in the nursery. Natural incubation conditions are superior to artificial conditions and it is suggested that eggs in the nest can therefore tolerate extended cooling far better than the nursery egg. Providing the cooling period in the nursery is less than 10 minutes and only occurs once a day, then hatchability will not be affected, however, if the cooling period is long and more frequent, then the developing egg may well be compromised.

Before placing the egg in the incubator for the first time, candle it closely for damage. Any chips or cracks must be repaired before introducing the egg to the incubator. Details on repair methods are found in the section, *Repairing Damaged Eggs*. Any faeces or nesting material stuck to the shell should be removed with a tissue and warm saline solution.

INCUBATION

The three key elements that constitute successful incubation are temperature, humidity and turning. Establish these within acceptable parameters and nature will do the rest. The task is actually easier than it sounds with the technology now being applied to incubators. A quality thermostat will control the temperature to within ±0.2°C. Humidity is easily regulated (in fact, some of the more expensive units do this for you automatically) while auto-turning is smooth and regular. Provided that the majority of your chicks are emerging healthy and unaided, adjustments should be slight. Except where a worrying trend develops, never make radical alterations on the basis of one or two losses.

Dry Temperature

Most Australian aviculturists set the incubator dry temperature at 37.2°C, and this setting works well for most species. Acceptable extremes are 36.6°-37.7°C. Venture outside these limits and you will compromise the hatchability. Modern incubators with reasonable insulation materials, quality thermostats and a fan should fluctuate no more than 0.2°C either side of the setting. Having said that, if the incubation room itself is subject to considerable temperature variations during a 24 hour day this will slightly affect temperatures within.

Eggs that are incubated at cooler temperatures may hatch up to several days later than expected, while eggs incubated at warmer temperatures tend to hatch a little earlier. Should the chicks be consistently hatching on either side of a typical hatch period, and suffer also from hatching problems, it is possible that the temperature may be too far from the optimum setting.

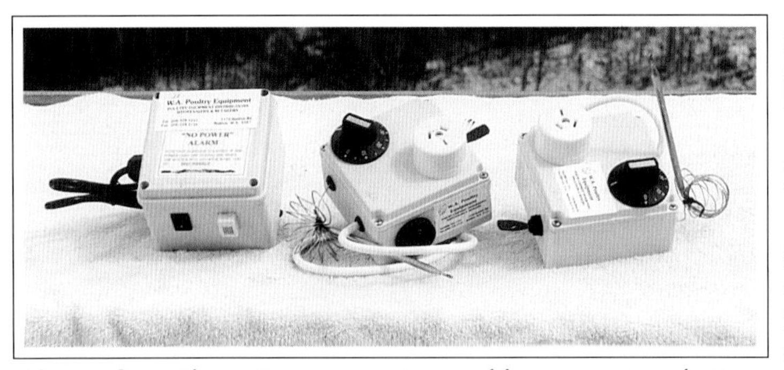

Above: Once the units are running and have eggs incubating, two control units can be of assistance. A No Power alarm will give off an alarm if the power goes off, day or night. The Hi-temp Control Unit can be set up to give off an alarm should the incubator/brooder temperature rise to dangerous levels.

An unlikely but possible cause of this problem could be a faulty thermometer.

Checking Your Thermometer

Before leaping into the exciting world of artificial incubation it pays to check the accuracy of your thermometer. This is done simply by comparing it with another over a day or so. If there is a difference of more than 0.5°C, it will be necessary to establish which thermometer is at fault. Consider the fact that your incubator may contain varying temperature zones, so check it carefully by placing the thermometer in various spots within the incubator, noting any significant differences. In many still air incubators and in some fan forced units, depending on the shape, there may be areas that are unacceptably warmer or cooler. Check and record the temperature once a day, to ensure that everything is on track.

Humidity

As an egg develops it loses fluid and therefore weight. The rate of this fluid loss is controlled by the level of humidity within the incubator. The more moisture in the air, the less the egg will lose - and *vice versa*.

Humidity is directly linked to the water surface area. Increasing the surface area raises humidity, decreasing it, lowers humidity. Should the humidity rise too far above the desired setting, there are several ways of lowering it. Firstly, the air vents may be opened a little further. If this is not enough, the actual surface area of the water needs to be reduced. This can be done by emptying one of the water jackets or by floating pieces of foam on top of the water, effectively reducing the surface area. If the incubator is a basic model, in which you place your own water containers within the unit, simply use smaller dishes or remove one. In extreme cases, where all water has been removed and the humidity is still too high, it may actually be necessary to use a dehumidifier to remove moisture directly from the air, however this situation very rarely arises, in Australia.

A humidity or wet bulb thermometer reading of 82°-83°F is the preferred setting by most aviculturists for incubating the majority of parrot species. Some report excellent hatch rates with a setting of 84°-85°F, while one breeder claims to do well at 79°-80°F. However, as a rule, the humidity reading should not move outside 82°-83°F wet bulb for more than a day or so, as this could prove detrimental to the developing embryo.

Turning

There is much debate over how often the parrot egg should be turned during incubation, and while opinions vary, all agree that the minimum requirement is three times a day. The purpose of regular turning is to prevent the contents from sticking to the inner membrane, and as the hen does this in the nest, it must also be performed in the nursery. Many auto-turn units turn the egg anywhere between 12-48 times during a 24 hour period. This frequency produces an excellent hatch rate. Some aviculturists also elect to hand-turn the egg 180 degrees once a day, to ensure that it makes a full revolution. This can certainly do no harm and may, in fact, be of some as yet unrealised benefit. Many others simply leave the machine to do the work and still enjoy excellent hatch rates. Should you choose to perform this additional turn by hand, it can be done while you candle the eggs, daily. Monitor the turning action of the auto-turn machine, as the moving parts can become dirty and dry, producing a jerky rather than smooth action.

If you are hand-turning, too many turns per day can become time consuming, therefore, eggs tend to be hand-turned less than when turned by an automatic unit. Pearl Coast Zoo had a very successful hatch rate with their hand-turned eggs by turning the eggs 180 degrees, four times a day. This was performed at 7.00am, 12.30pm, 4.30pm and 9.00pm. Hand-turning should consist of a gentle roll or turn without jerking or jarring. A pencil mark somewhere on the egg helps to determine just how much it has been turned each time. Turn the egg the way it would naturally roll, not end over end.

Turning must begin the day the egg is placed in the incubator and continue until the beginning of the hatching process. When using an auto-turn unit, it pays to note the position of the turning bars whenever you are in the room. The reason for this is that auto-turn mechanisms *do* occasionally break down, and may go unnoticed for quite some time. During incubation, the egg should be placed on a flat surface, as this will enable it to assume a natural position, in which the end of the egg containing the air cell is slightly elevated above the horizontal plane. Nature designed the egg to sit this way and this should be duplicated in the artificial environment. In units where the egg is rolled across a grid via bars, the egg rests naturally in the normal position. Machines that clamp the egg between rails also allow the egg to sit this way. If the egg is being hand-turned, it can be placed on a piece of gauze either directly on the base or on a simple wire platform. Wherever and however the egg is placed, it should not be in a position where it can rock around.

Fertility

At this point, you should have a reliable unit, a freshly laid egg being regularly turned and accurate heat and humidity settings. Incubation is now under way! Fertility can be confirmed on Day 5 or Day 6, and sometimes as early as Day 3 with a good candler and a little experience. Candling at any stage (but more so at this point) is best performed in a dark room, where the light will better reveal the internal features.

The first real signs of fertility will be the appearance of faint blood vessels around the centre of the yolk. Within another 24 hours, more vessels will be noticed and in another two to three days, a network of blood vessels will begin to radiate throughout the yolk from the embryo. In larger eggs, the embryo's heart can often be seen pumping from Day 8 onwards. Exercise patience, and never discard an egg until infertility is indisputable. Occasionally, for reasons not yet understood, an egg may have a 'slow start', not showing visible signs of blood vessels until up to Day 7.

Early death of the embryo will result in all the blood vessels receding to the fringe of the yolk, forming what is termed a *blood ring*. This should not be confused with the very first signs of fertility, which may actually be a ring of blood around the centre of the yolk with faint vessels connecting to the germinal disc. Daily candling of the egg is strongly recommended, particularly during the early stages. It is only by becoming familiar with the way an egg naturally develops that you will be able to recognise a potential problem, during which time the egg may still be saved.

Development Phase

Around Days 8-10, the vascular system begins to extend around the entire egg and the fluid section begins to darken. By mid-term, many eggs are quite dark and details of the chick become difficult to distinguish, the only visible signs of life being the presence of a few large blood vessels close to the air-cell line. Occasionally the chick can be seen moving as early as halfway through the full incubation period. At this point the air-cell becomes the best indicator of progress. Many experienced breeders can accurately predict the age of an egg simply by the size of the air-cell. Generally, the air-cell expands uniformly down the egg - but do not be alarmed if the air-cell line tilts as incubation proceeds. This will happen in some normally developing eggs.

As in the fertility phase, never discard an egg unless you are absolutely certain that the chick has died. Many eggs have developed dark spots or other unusual features,

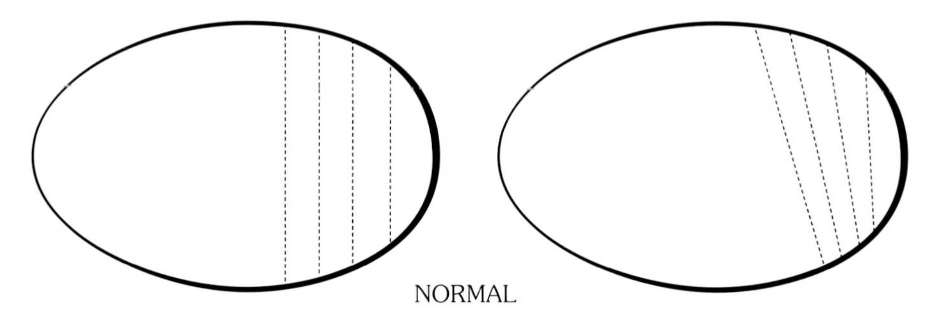

NORMAL

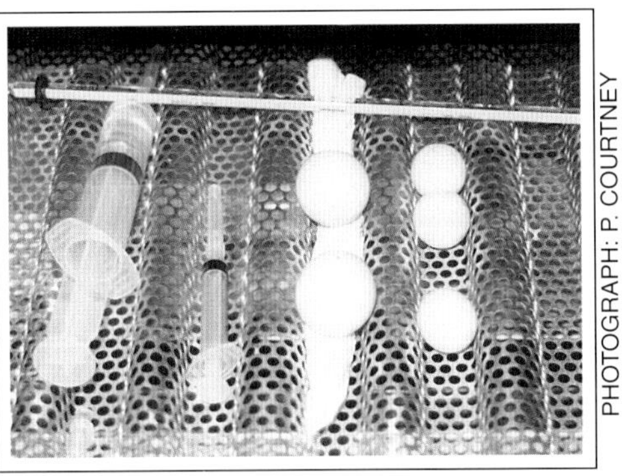

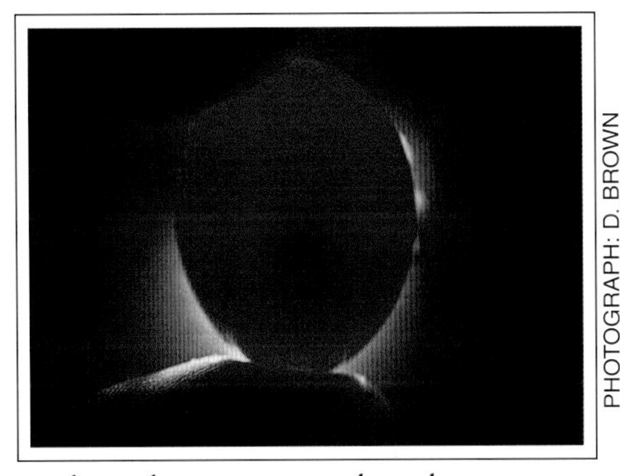

Above left: The eggs are hand-turned every three hours up to drawdown stage. Syringes and contents being warmed ready to feed chicks that hatched earlier.
Above right: An egg at 7 week incubation showing early blood vessels formed and central embryo.

including a complete absence of blood vessels in an area of the egg - and yet, they still hatch successfully. At any rate, there is nothing that can be done about internal aberrations except to continue the incubation and hope that the chick begins to hatch - because at *that* point, there is plenty that can be done.

There are two indications of embryo death. The first is the receding of active blood vessels around the air-cell line. Although the egg has darkened as it has developed, these cherry red vessels close to the air-cell should have remained visible. Once the chick dies, they turn brown, and many recede as the chick becomes, a dark blob in the centre of the egg. The second is the colour of the egg - from a solid glossy look, it starts to assume a dull, chalky grey appearance.

An interesting phenomenon, not recorded in other literature on this subject, has been noticed by several breeders incubating Black Cockatoo eggs. Somewhere around Day 13 and Day 14 of development, the egg contents turn washy and milky and it appears as if the whole egg has been heavily shaken. Movement of the egg will cause the contents to wobble around and to all intents and purposes the yolk appears to have collapsed. Black Cockatoo eggs have been presumed dead at this stage and are sometimes discarded. However, such eggs pass through this stage and resume normal appearance within a couple of days. They then proceed to hatch successfully.

Wherever possible, allow eggs to receive at least a week of natural incubation before pulling, even if this means placing them under a foster hen. Natural conditions will always be slightly superior to artificial, and the more natural incubation the egg receives, the higher the hatchability. There is a higher incidence of malpositioning and hatching problems in eggs artificially incubated from day one as compared with eggs left under a hen for two weeks or more. Also, the longer the period of natural incubation, the more the egg will tolerate marginal conditions once it is in the incubator.

HATCHING DETAILS

Once the egg begins the hatching process it is on the home straight, however, hatch time itself is a crucial period and requires regular monitoring. The air-cell now becomes the centre of attention until the first sign of external pip is observed. The over-anxious breeder may intervene in the hatching process too early, while others are too hesitant to assist a chick that in fact, needs help. Until experienced, the line between a chick doing well and a chick in need of help will be a little obscure. Generally speaking, however, the vast majority of eggs hatch unaided, providing the settings were within acceptable levels and more so, if the egg received some natural incubation initially.

Records

Before discussing the finer details of the hatching process, the importance of keeping records from herein, needs to be emphasised. The hatching process is much more than a chick simply poking a hole in the shell and crawling out - it is a progression, of which every step needs to be monitored and recorded. Plot ahead in your daily incubation records, marking five days in advance of expected hatch date as the time to expect earliest hatching signs.

Drawdown

The first indication of the hatching process is the noticeable movement of the air-cell, referred to as *drawdown*. The best way to monitor drawdown is to put a pencil line around the air-cell line on Day 6 prior to the expected egg hatch date. Candling on Days 5 and 4 should reveal movement of the air-cell down one side of the egg. If the air-cell was tilted during development, this tilt becomes exaggerated, however, if the air-cell was pulling down evenly, it will now develop a tilt. Once movement of the air-cell is detected in a 24 hour period, the egg should be candled at least morning and night, to ensure close monitoring of its progress. Each time the egg is candled, pencil in the new position of the air-cell. At this point the egg will be roughly four days away from hatching. The air-cell, though drawing down, will still be in a uniform line as shown in the diagrams.

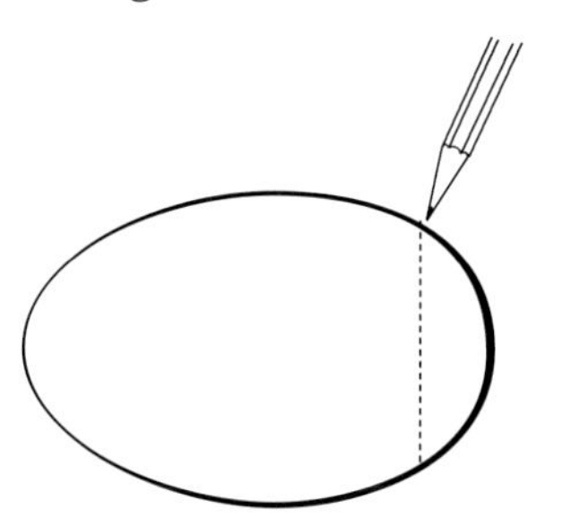

Day 6 - Pencil in air-cell line

Day 5-4 - Noticeable tilting

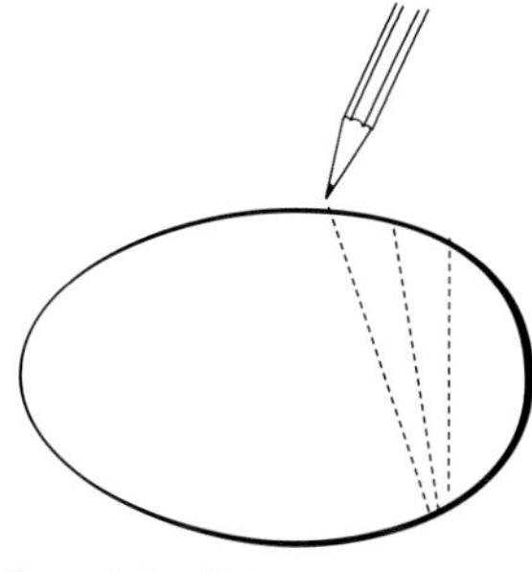

Day 4-3 - Tilting increases

Internal Pip

The next step of the hatching process, *internal pip*, will now occur. The chick will pip through the air-cell membrane and begin to enter the air-cell using a sharp ridge on its top beak called the *egg tooth*. It is at this point that the egg should be transferred to the hatcher, where it is no longer turned and the humidity is elevated. Internal pip is recognised by a break or 'umbrella dip' in the air-cell line.

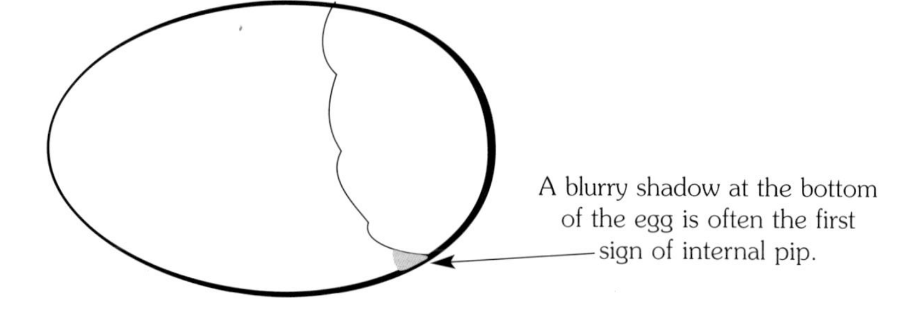

Occasionally, there will be no discernable break in the air-cell line before external pip begins. Different texts describe the drawdown/internal pip stages in various ways, which may lead the novice to ask, 'At exactly what point must I stop turning the eggs and place them in the hatcher?'

Drawdown generally begins at around Day 5 prior to expected hatch date and this tilting of the air-cell continues for up to 48 hours. This means the chick is positioning itself to shortly move into the air-cell. It is fine to keep turning the egg while this is happening, but as soon as the first sign of a break in the air-cell line is noted, cease the turning process and move the egg to the hatcher. To keep turning at this point will only confuse the chick as it attempts to begin external pip. Often the first sign of internal pip will be seen at the bottom of the egg, where the air-cell has been 'hinging'. In this area the air-cell line will appear blurry, and often, movement in the form of flickering shadows, will be seen as the chick's body protrudes into the air-cell.

Within 12-24 hours of the first signs of internal pip, the air-cell will begin to collapse visibly, as the chick enters further into the air-cell and begins working on the shell. If you are uncertain as to exactly when to cease turning, take the egg off auto-turn on Day 5 prior to expected hatch date. With the air-cell positioned as shown in the following diagram, simply hand-turn the egg 45 degrees either way, three or four times a day, until the air-cell collapses or external pip appears.

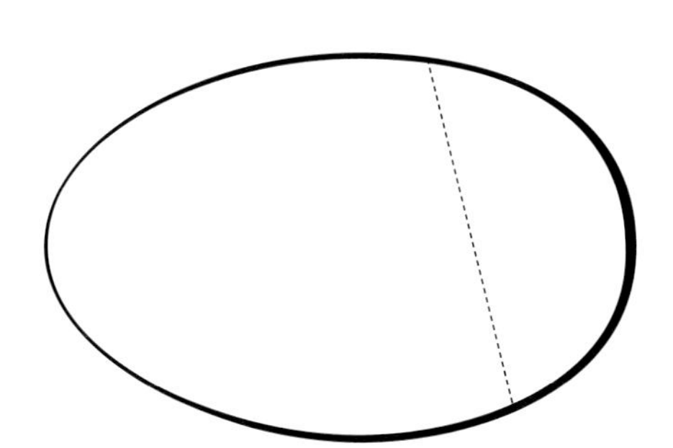

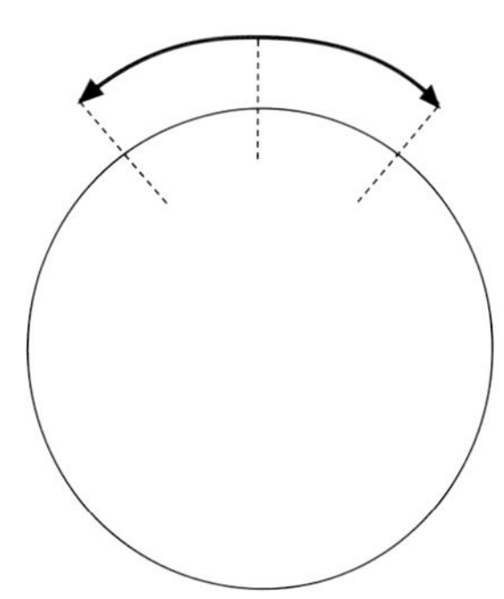

Hand-turning method

This way the contents will not stick and the turning will not be aggressive enough to confuse the chick as it moves towards the top of the air-cell and begins external pipping. If at any time, while candling an egg during the hatch process, you hear vocalisation from the chick, or feel tapping from inside, then it is obvious that you have missed drawdown and the chick has already broken into the air-cell (internally pipped).

Into the Hatcher

Confirmation of internal pip is either by a break in the sharp air-cell line or by vocalisation and tapping. Once internal pip is confirmed, which is generally three or four days prior to hatching day, place the egg in the hatcher. Place the egg in a dish or

PHOTOGRAPHS: D. ANDERSEN

Left: External pip.
Centre: Chick breaking out of the shell.
Right: Chick hatched ready for the brooder.

Left: White-tailed Black Cockatoo chick, hatching.
Below: Hole punched in the shell by a chick about to hatch.

bowl lined with tissues and padded in such a way as to prevent the egg from rolling. Take care not to pack around the egg so tightly that the chick cannot kick free from the shell. The container should be twice the depth of the egg, as some chicks are surprisingly agile once hatched and if they fall out of the container, they may injure themselves on the floor of the hatcher.

Humidity in the hatcher should be raised to 94°F wet bulb or higher to prevent the membranes from drying out and trapping the chick. As already mentioned, chicks *have* been hatched in incubators at a humidity of 82°-83°F wet bulb, but this is not recommended. Hatching is hard work and the chick will need all the help it can get. Depending on the model of your hatcher, the filling of all the water jackets may not raise the humidity sufficiently. Should that be the case, there are a number of remedies.

Firstly, place lids or small trays of water within the hatcher. By increasing the surface area of water you will guarantee a rise in humidity. Alternatively, drape strips of paper towel between and around the water jackets and have part of the towel hanging in the water. This way the entire piece of towel will absorb water and greatly increase the rate of evaporation. Closing the vents a little also helps to raise the humidity slightly, however, take care to leave them at least partly open. Never close the vents altogether.

At this stage of hatching the dry temperature can be lowered slightly, although not lower than 36.9°C. While some aviculturists choose to lower the temperature, others do not, leaving it at the same level as the incubation temperature.

External Pip

Once internal pip has occurred and the egg has been placed in the hatcher, anticipate the next stage of hatching - *external pip*. All species vary, as do individual chicks, however the first signs of external pip should appear within 24-48 hours after internal pip or collapsing of the air-cell. Occasionally it will be as soon as 12 hours after internal pip, which simply means that the chick is a fast worker, or has moved

Right: Egg tooth on a newly hatched White-tailed Black Cockatoo.

into the air-cell unnoticed and earlier than expected. By candling the egg two or three times a day at this stage, external pip will be detected early and you will have some idea of when the actual hatching is likely to occur.

Above: On hatching, clean up the chick, remove membrane stuck to chick and place in the brooder to dry out. Note egg tooth.

External pip will first be noticed as a small, slightly raised piece of shell with fine cracks forming a star shape, the result of the chick tapping away at the shell with its egg tooth. Generally, this will be within the air-cell, at the top of the egg. When candling at this stage it is not uncommon to feel the chick tapping from the inside, an excellent sign that the chick will probably hatch unaided. More star-shaped cracks will appear during the next 24 hours, before things go quiet at the original pip site, while the chick rotates inside the egg working away at the remainder of the shell circle. It then returns to the original pip site and continues to work away in preparation for kicking free of the egg.

When a flap of shell has actually folded back and a hole in the shell is visible, prepare for the chick. If this happens late at night, it may be wise to set your alarm clock for some time early morning to ensure that you will be present to examine the newly arrived chick. Some chicks hatch within a few hours of creating that first hole in the shell and it is important to be present as soon as possible after hatch for several reasons.

The chick may get stuck in one of the shell halves or the yolk sac may not be totally absorbed, and therefore in need of attention. Alternatively, it may be a wet hatch, where the chick is caught up in a tangle of faeces or fluid - or perhaps a piece of membrane is stuck over the nostrils of the chick.

Should the initial pip site remain unchanged for 48 hours after the first pip cracks appeared, the chick may be in need of attention, as dealt with in the section *Overdue to Hatch*. However, the vast majority of correctly incubated eggs that begin hatching, complete the process unaided. If the external pipping marks appear in any portion of the fluid section of the egg, then it is obvious that the chick is malpositioned, as dealt with in *Troubleshooting*.

It's Hatched!

Between 24-72 hours after external pip there should be a healthy parrot chick in the bowl, and now it is simply a matter of cleaning the chick. Any pieces of membrane stuck to the chick should be removed before they dry out. Using Betadine™ diluted 1:10 with sterile water, treat the navel area using cotton wool buds or a small paintbrush. This is an important procedure, because until the umbilicus is properly pinched shut and sealed with skin, it remains an avenue by which infection can enter the chick's body. Perform the cleanup under a heat source to prevent the chick chilling. Once completed, the chick should be placed in the brooder set at a temperature of 36.6°C to dry out before the first feed.

PROBLEMS AND TROUBLESHOOTING

Following is a list of the more common problems you may encounter in artificial incubation and hatching of parrot eggs. Do not be alarmed by the number of potential problems, nor be concerned by the proportion of text devoted to this area. Potential problems need to be discussed in more detail so that they can be dealt with quickly and efficiently.

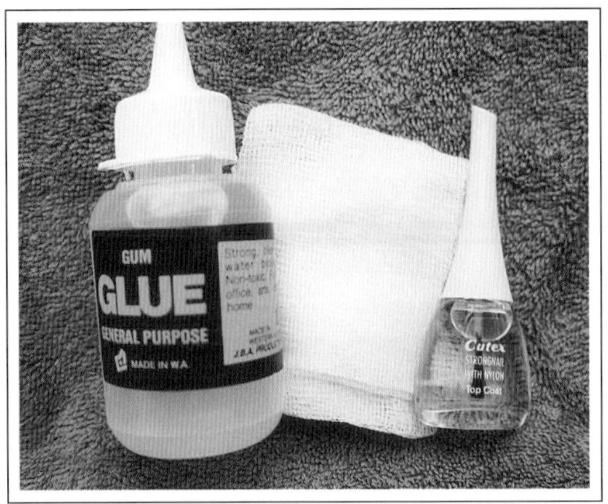

Glue and sterile gauze used for repairing damaged eggs. PVA woodworking glue is also used by some. Others successfully use clear nail polish.

Most of the following problems fall into one of two categories:

- Operator error/marginal incubation practices
- Very occasional inherent genetic defects

Chicks in this second group have a much greater chance of hatching in the nursery where assistance can be offered.

The most important point to remember is that you must never alter your settings or your regular routine unless there is a recurring problem.

Damaged Eggs

All eggs need to be candled to check for damage before they are placed in the incubator. Often, eggs laid on the ground or pulled from beneath a rough hen, are damaged or cracked. Unless they are clearly beyond repair, do not be hasty in discarding them, as there is nothing to lose by repairing an egg and placing it in the incubator.

Cracks, dents or holes in the shell will alter the evaporation rate of the egg to a certain extent and offer easy access to bacteria and fungal agents. For these reasons, repair is important. Suggested ideal water soluble glues for repairing eggs are PVA woodworking glue or clear gum. With a straightforward crack it is simply a matter of running glue the entire length of the crack to seal it. The glue will totally block respiration through any area it covers, so keep it centered on the crack as much as possible. Do not allow the excess to run down either side of the egg. Small dents or 'cave-ins' can be repaired in much the same way, however, where the damaged area is large then it becomes necessary to strengthen it with gauze, which should be bonded with the glue. Apply the glue directly around the perimeter of the hole, then place a piece of sterile gauze just larger than the hole, over the site. Allow the glue to dry before repeating the procedure. Seal the gauze with glue after the final layer. The greater the damage to the shell the greater the likelihood of early death, however many eggs considered 'written off' have been repaired and produced a chick. Where the inner membrane has been pierced or torn and there is albumen appearing at the hole then hatchability is definitely compromised, however it is worth attempting repair. Where there is a repaired area over the air-cell, keep in mind that the chick may begin external pipping under that exact spot. Should this be the case, it becomes necessary to remove the gauze and glue, once the chick has begun internal pipping and entered the air-cell. As the glue is water soluble, it is simply a matter of moistening the egg using a small paintbrush and sterile water, before removing the gauze.

PHOTOGRAPH: D. BROWN

Left: An egg with early embryonic death showing indistinct embryo and faint 'blood ring' at periphery of yolk.

Early Death of the Embryo

Death of the embryo at an early stage will be indicated by the receding of the blood vessels within the yolk, to the fringe, where they will form a blood ring. Causes for this early death can be several, such as rough handling at the time of pulling or poor turning during incubation. If using an auto-turn unit, regularly monitor the quality of the auto-turn mechanism and make certain that the roll is smooth and gentle. Periodically check the machine for excessive vibrations, as small embryos are particularly sensitive to these.

Extremes of temperature or humidity, particularly over a long period of time, may kill an embryo. Accurate records and constant checking of instruments will eliminate this possibility. Check for varied temperature zones within the unit, as some eggs may be in a safe zone, others in a marginal zone. Should deaths regularly occur in eggs that came from the same parents, it is possible that there is a genetic factor involved, and if all other possibilities have been eliminated, this may be the case.

Death during Development

While radical extremes of humidity and temperature will bring early death, marginal settings will often cause death later in development, as the embryo slowly succumbs to the poor environmental conditions. This is where quality instruments become important, as, in the absence of the operator, unacceptable fluctuations can be recorded. If more than one egg dies within a short space of time, consider the possibility of bacterial or fungal invasion. This problem can be largely eliminated by washing your hands carefully and regular disinfection of incubator/hatcher.

Overdue for Drawdown

Having planned ahead on your record sheets, you will have estimated the date on which the egg is due to hatch. Deduct five days from this hatch date and you will have the drawdown date. You may be concerned if there is no noticeable movement of the air-cell around this time. Remember that while there are specific incubation periods given for different species, not all eggs stick to dates. The temperature of the incubator will play a part in the hatch date. A lower temperature causes the egg to hatch later and *vice versa*. It is possible for an egg to successfully hatch two to three days late. Drawdown cannot be induced, so patience is the key.

Overdue for External Pip

Between 12-48 hours after the air-cell has begun to pull away or dip in places (internal pip), expect to find signs of external pipping somewhere in the upper air-cell. Candle the egg carefully during this stage, as external pipping can be easily missed, causing unnecessary anxiety. Look for pip marks in the solid section of the egg, indicating malpositioning of the chick. While waiting for the external pip to appear, keep an eye out for two positive signs; vocalisation from the chick (indicating that it has broken through the air-cell and is now breathing air) and occasional tapping from within the shell, which can be felt if the egg is held in a bare hand.

If 48 hours elapse after internal pip and there is still no sign of external pip, then a small hole should be made over the air-cell, to allow oxygen to reach the chick. There is no harm in doing this, as by now the chick would have punched its own hole through the shell.

To make a hole, twist a sterile syringe needle into the very end of the air-cell. Move slowly, so as not to break through the shell too quickly and puncture the chick. Having done this, replace the egg in the hatcher, ensure that the humidity is 94°F wet bulb or higher, and monitor the egg for another 24 hours. If the chick has still not externally pipped at this point, it is suggested that the chick is either too weak to hatch or in the wrong position to do so. It is now time to chip away a hole over the air-cell in order to take a look at what is happening inside.

Place the egg under a heat source and using tweezers, make a hole as indicated in the diagram.

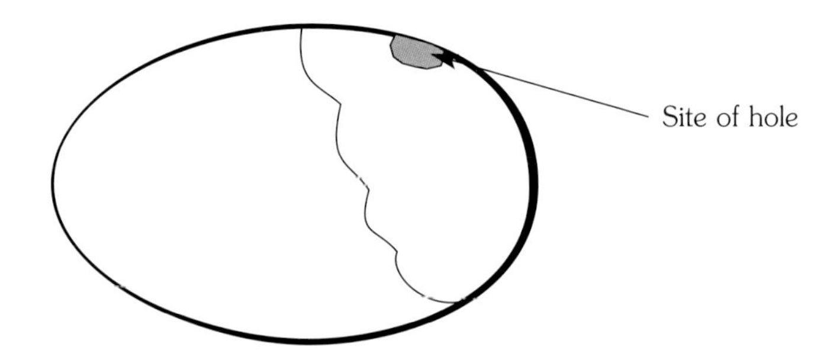

This is a simple procedure, as there are no blood vessels over the air-cell to be concerned with. The hole needs to be large enough to allow clear viewing of the inner membrane. In the case of a Black Cockatoo egg, for example, a hole 15mm in diameter is adequate. Regardless of whether the chick has penetrated the inner membrane with its egg tooth, the same procedure applies from herein - using a sterile 1ml syringe, drop one or two drops of sterile water onto the membrane covering the chick. This will highlight any active blood vessels still present in the membrane. The blood vessels will appear as pink-red lines across the membrane. Whenever a hatch is being assisted, it is unwise to proceed while active blood vessels are present. The egg must be returned to the hatcher and the procedure repeated every six to eight hours until the vessels recede. As soon as the blood vessels have receded, use sterilised tweezers to pull back the membrane, beginning at the site where the chick's egg tooth penetrated the membrane.

This should be performed as slowly and carefully as possible. Having exposed the chick's head and beak, return the egg to the hatcher monitoring closely. Never be in a hurry to pull a chick from the egg, unless the situation is urgent, as the result may be an insufficiently absorbed yolk sac. Once the chick has been revealed, there should be some progress in the next 12 hours, but if not, move on to the procedure outlined in the following section.

Overdue Hatch

Once external pip has occurred, expect a chick in the next 24-72 hours. After initial work around the pip site, things will go quiet as the chick rotates inside the egg, working on the shell circle. Then, close to hatching time, activity should resume at the initial pip site and more star-shaped cracks, then holes, will appear.

Vocalisation is an excellent indicator of the chick's condition around this time. If it sounds strong and its movements are not erratic, then all is probably well. If the call sounds weak or panicky, however, the chick is most likely weak and in serious trouble. If there is no noticeable progress 48 hours after the initial external pip, it may be wise to enter the egg.

Once the decision is made to enter the egg, your actions can afford to become a little more aggressive than before external pip. Chip away a large enough hole over the air-cell to allow access to the chick and using sterile water, check the inner membrane for active vessels. Providing there are none present (and at this stage there shouldn't be), peel back the membrane and allow the chick to fall into the air-cell. At this point, as the chick is now occupying the whole egg, it is relatively easy to check the navel area. If the yolk sac is unabsorbed, return the egg to the hatcher and check later. If it is sufficiently absorbed, break away more of the shell and return the egg to the hatcher, where the chick will then proceed to make its own way out of the shell. If, however, the chick appears extremely weak or dehydrated, it will need to be assisted out of the shell and fed electrolytes immediately.

Malpositioned Chick

During candling for external pip, you will also be looking for pip marks over the solid section of the egg, indicating a malpositioned chick. Sometimes a chick will pip right on the air-cell line and as hatching proceeds the pipping moves further into the air-cell portion of the egg, in which case intervention is unnecessary. If however, the chick is pipping further into the solid section of the egg or the pip appeared in the small end to begin with, then the chick will need help.

Thankfully, malpositioned chicks almost invariably pip at the top of the egg. The problem presents itself as the chick begins to rotate and work its way around the egg, if it *can* rotate which many malpositioned chicks cannot. Drowning becomes a possibility as the chick moves down the egg and although there is not a lot of fluid left in a correctly incubated egg at this stage, it takes very little to kill a chick. Intervention should be immediate if pip marks appear in the small end of the egg. Very carefully peel back the slightly raised flaps of shell and drop sterile water onto the membrane to check for active blood vessels. If vessels are present, return the egg to the hatcher, this time sitting the egg with the small end raised, with a pin hole in the air-cell end to relieve the internal pressure and create more room for the chick. Check the egg again in a few hours and as soon as the blood vessels have dried, begin exposing the chick. Proceed cautiously, as a burst blood vessel could result in the chick bleeding to death. Concentrate your efforts on uncovering as much of the chick as possible without actually allowing the chick to leave the shell prematurely. Once the egg has had most of the small end removed, many chicks stop trying to rotate and simply rest before eventually struggling free of the egg.

The cause of malpositioning in individual eggs may never be determined conclusively, but improper turning has been indicated as a possible factor. However, if the egg *was* turned properly, it may well be a natural aberration - after all, it does happen in the nest where all conditions are, theoretically, optimum. Whatever the reason for their plight, malpositioned chicks have a greater chance of surviving the hatch in the nursery than in the aviary nest.

Unabsorbed Yolk Sac

During the hatching process the yolk sac is absorbed into the chick's abdomen and the skin closes over it. The first area to check once the bird emerges is the navel, which will indicate whether the yolk sac has been absorbed. Once you have developed a successful system of management, the vast majority of chicks will hatch without problems - but occasionally, a chick will hatch with the yolk sac unabsorbed. This is generally the result of either the humidity or the dry temperature being too high. In the case of an assisted hatch, it could mean that the chick was pulled from the egg too early.

If the yolk sac is largely drawn in, with little exposed, it is best to treat the area with Betadine™ solution and place the chick in the brooder. Continue to treat the area with Betadine™ for one to two days. During the next two or three days the navel should begin to close up and pinch off the extended piece of sac, which will then dry up and fall off. In a case where the yolk sac is far from absorbed or 'hanging out' of the chick's abdomen, veterinary assistance should be sought immediately. At times a chick will hatch with an umbilical cord of membrane attached to its navel, despite the fact that the skin has closed over the yolk sac. A few days in the brooder will see this dry up and fall off.

Wet or Dry Hatch

The factor that determines whether a hatch is excessively wet or dry is the humidity. Should there be a trend of chicks hatching too wet, the humidity needs to be lowered and *vice versa*. A day or so of incorrect humidity will not produce a wet or dry hatch, however, more extended periods will begin to influence the end result. For example, if the hatcher ran out of water for a day or the weather was excessively humid for 24-48 hours, these conditions will be tolerated by an otherwise on-track egg, but are by no

means desirable.

A wet hatch occurs when a chick hatches from an egg containing excessive amounts of fluid. Often the chick is too weak to hatch unaided and will need assistance. Many chicks actually drown during hatching if this problem is not quickly diagnosed. The chick will appear swollen and bloated as a result of retaining too much fluid within the tissue. Both the egg and chick will be somewhat slimy. As a rule, providing there was no aspiration during the hatching process, these chicks recover in two or three days and proceed to develop into fine birds.

Above: Assisting a young Eclectus to hatch. This bird was actually fed in the shell while being assisted to hatch.
Below: Young Eclectus chick having been just assisted to break out of the shell. (Photographs - P. Courtney)

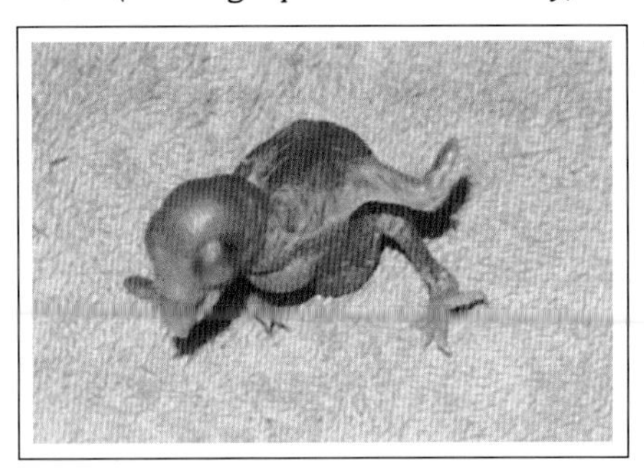

A dry hatch can be recognised by just that - an egg that is very dry on the inside upon hatch and a chick that has a wrinkled, shrunken appearance. Often overly dry membranes will stick to the chick during hatch and loud vocalisation with little progress will alert you to this problem. Dry hatch is the result of unacceptably low humidity during incubation. The minute such chicks hatch they should be fed fluids (lactated ringers - Hartmann's solution) immediately.

The other factor that determines a wet or dry hatch is the shell thickness. Occasionally the egg shell will be considerably thicker or thinner than usual, which will alter the evaporation rate of the egg. The experienced aviculturist will recognise whether or not an egg's weight loss is on target simply by checking the size of the air-cell at any time during incubation. An egg that still has a small air-cell halfway through incubation has not lost enough fluid, while if the air-cell is expanding too quickly, it is losing too much. Either way, adjustment of humidity settings will be needed to put such eggs back on track. Providing the majority of your eggs are hatching unassisted, the humidity will only require minor adjustment, if any. Eggs that are not losing weight at an appropriate rate, are those that benefit the most from egg weighing.

EGG WEIGHING

The ideal weight loss for the parrot egg between Day 1 and external pip is 15% to 17% of its original weight when laid. The recommended dry and wet temperature settings previously discussed have proven to be so effective in mainstream incubation that individual monitoring of weight loss is rarely practised, even in larger nurseries. However, weighing is undoubtedly a more precise method of achieving the best possible conditions at hatch and is particularly useful when an egg is not developing normally. It is not the intention of this book to deal in detail with such a specific aspect of incubation as egg weighing, albeit a little information on the subject is appropriate.

To accurately weigh eggs and thereby ascertain their weight loss, you will need a set of electronic scales that measure to two decimal places, especially when dealing with eggs from the smaller species. Weigh the egg on Day 1, then subtract 15% to 17% of that weight. This will give the desired weight of the egg at external pip. That 15% to 17% of the freshly laid egg now needs to be divided by the number of days between zero and external pip (incubation period minus 2 days) to calculate the desired daily weight loss. Following is an example using a freshly laid Red-tailed Black Cockatoo egg weighing 25 grams, with an incubation period of 28 days.

25 grams X 16% = 4 grams (Total weight loss necessary over the 26 days between zero and external pip)
4 grams ÷ 26 days = 0.15 grams daily weight loss

Once the daily weight loss is established, plot ahead in your records the ideal daily weight of the egg. As the egg develops, record the actual weight of the egg each day alongside the desired weight. By doing this and assessing the overall progress every few days, the humidity can be altered if necessary, before the egg goes too far off track.

Eggs that benefit most from the weighing procedure are the occasional ones that are unusually thick or thin shelled, therefore in need of individual treatment, particularly with regard to humidity. This is why it is so important to have a second incubator if egg weighing. However, in the absence of a separate unit for individual treatment, there are other methods. An egg losing too much fluid (thin shelled) can have small areas of its surface coated with gum glue to slow the moisture loss, while eggs not losing enough weight can be sanded lightly with very fine sandpaper. The experienced breeder will not necessarily need to weigh the eggs in order to find out whether they are losing too much or not enough fluid, as the size of the air-cell at any particular age will be sufficient indication. As with most things in life, it is surprising what can be done with a little experience.

PUBLISHERS NOTE

It must be stated before beginning this section that there are many and varied opinions on handraising methodology. It is not the intention of this title to attempt to dissuade any breeder from his or her successful handraising regime. If it is working for you, don't change it. However, the following information is presented to inform and enlighten the handrearer on the methods successfully employed by the author. The primary principles and objectives set out are in essence the building blocks of all handraising methods.

As is mentioned by the author and strongly suggested by the publisher, communicate with other breeders, study their methods and regime and ask questions.

Ultimately it is up to the reader to decide on the most comfortable and suitable handraising method to employ. However, I have no doubt that the following information will greatly assist in the understanding and knowledge of handraising parrots.

Nigel Steele-Boyce
Publisher

Handraising Parrots

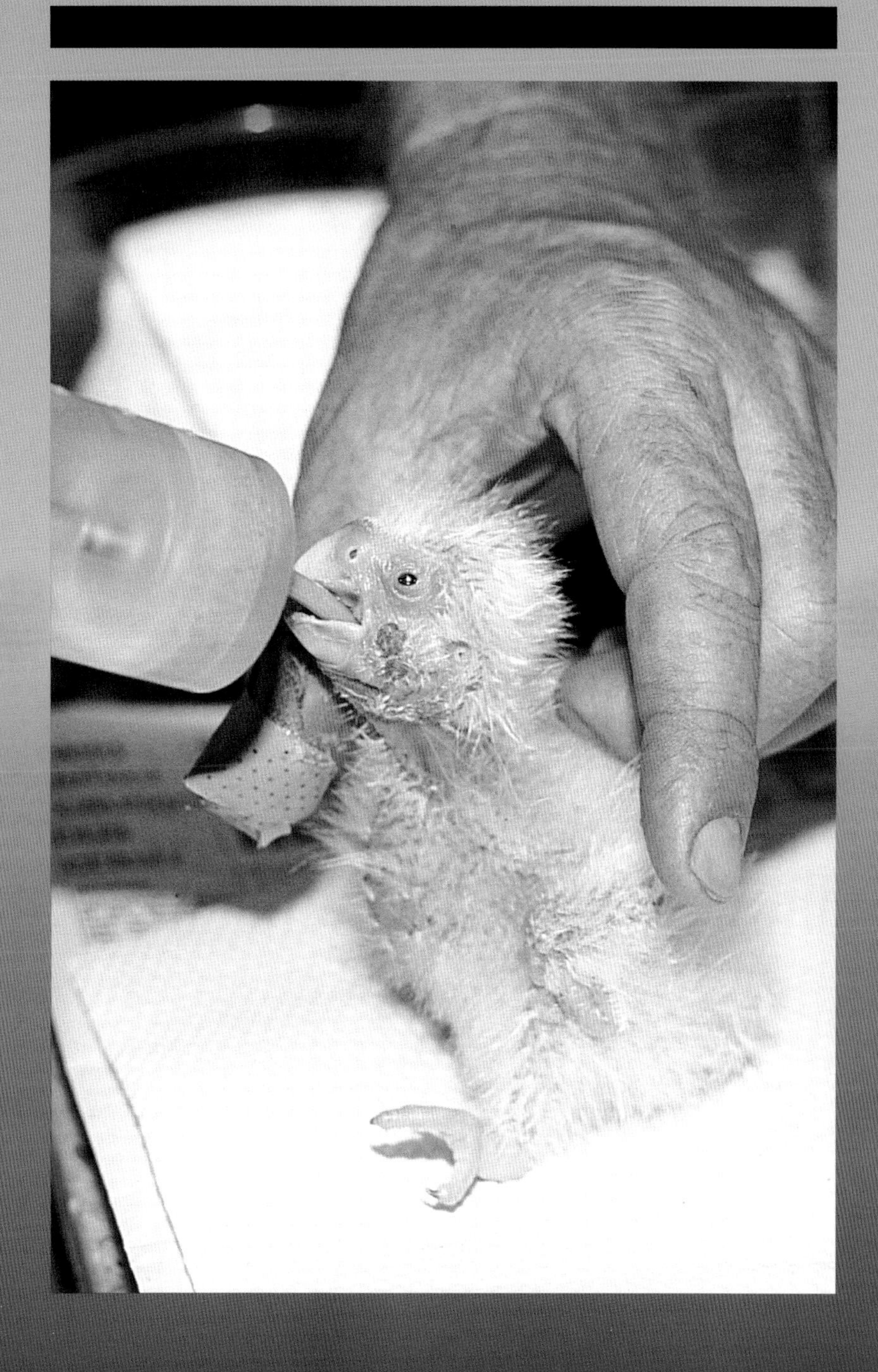

INTRODUCTION

At some stage of their avicultural life almost all parrot breeders will be confronted by the need or the desire to handraise a parrot chick. Whether it be due to the loss of one of the parents, a chick being left behind in a large clutch, or simply the urge to have a close family pet, handraising is an integral part of successful parrot aviculture; saving lives and increasing production.

Quality handraising needs to be clarified. It is one thing to pull a half-grown chick from the nest and simply keep up the formula until it feathers and begins feeding itself. That is relatively easy and even a poor diet and questionable regime will generally see the chick through. It is something else to take a newly hatched chick and to nurture it feed by feed into a robust, well-feathered individual without problems and wean it in a reasonable time. This requires dedication, patience and that most important element, experience.

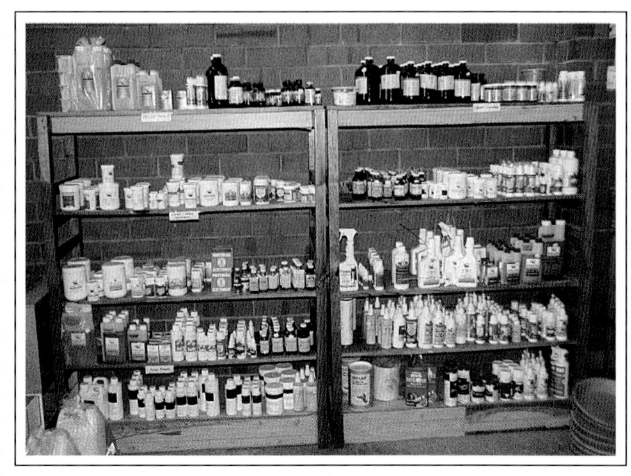

Above and below: We have everything at our fingertips to achieve the goal - to produce a chick comparable in size and stamina to that which nature would have produced.

Fortunately for aviculturists today, commercial formulas, state-of-the-art brooders, improved feeding techniques and the willingness of experienced aviculturists to share their knowledge have combined to take much of the hard work out of handraising, while lifting the quality of chicks produced, to a new level. However, 'the more things change the more they stay the same' and the overall goal still remains, to duplicate as closely as possible the natural growth of a parrot chick so that at the end of the process it is comparable in size and stamina to that which nature would have produced.

Will pay cash for single volume or entire collection.
Books considered Ornithology, Natural History, Aviculture, especially books on Parrots, Pigeons Poultry.
Send list with author, title, year published, publisher, description of condition and price for consideration.
Silvio Mattacchione & Co., 1793 Rosebank Road N., Pickering, Ontario, Canada. L1V 1P5.
Phone **(416) 831 1373.** Fax **(416) 831 3734**

MASK LOVEBIRDS. Albino and White Mask split Albino. Yellow and White Mask. Yellow Lime Fischers split Lutino. Lutino Fischers. Unrelated pairs. Phone Mike Nash, private breeder, Murwillumbah. Will freight. **(066) 722 925** (evenings)

BLUE RINGNECKS and splits. Lutino Ringnecks. Mature Pairs of Alexandrines. Surgically sexed, closed rung, unrelated pairs. Phone Mike Nash, private breeder, Murwillumbah. Will freight. Phone **(066) 722 925 (evenings)**.

BLUE RINGNECKS & Splits 1 to 3 yr old. Grey, albino, lutino. **ALEXANDRINES** S/S pr Plumheads, Slateyheads, 1 to 2 yr old. **ECLECTUS,** Young cocks & hens. Ph: Ian Adcock **(066) 721 646**.

SUN CONURES, cocks, hens unrelated prs S/S, closed rung. Hand & parent reared available. Will freight. Ph Ian Adcock **(066) 721 646**.

ECLECTUS, young unrelated pairs Phone: **(059) 522 484**.

AFRICAN LOVEBIRDS. Breeding White Yellow and Split masks - No Hybrids. Phone Les **(049) 516 993** evenings.

INDIAN RINGNECKS. Blue Ringnecks, Double split & Green Split to Blue. For further details

PLUMHEADS S/S This seasons young Northern Rosellas and Blue Princess. Guaranteed quality. Parent reared from 100% Fertile clutches. Phone Alan on **(060) 218 030**

GOLDEN-SHOULDEREDS. Northern Rosellas, Rosa Bourkes plus splits. All quality birds. Phone **(042) 832 014** B/H or **(042) 836 196**.

ECLECTUS young hand reared birds. Micro chipped if desired. Phone **(043) 591 685.**

BIRD AVIARY - colourbond, homemade. 3 compartments. Flights suspended. 3.6m x 4m x 2.1m As new $1100 ono. Phone **(045) 721 800.**

COCKATIELS, Cinnamon, Pied & Pearl. Will Freight Ph: Ron Smith **(057) 821 1220 A/H**.

WHITE-FACED COCKATIELS & Splits. Specialising in breeding Cockatiels & their mutations Ph Mike Anderson **(07) 245 1709**.

ECLECTUS, young hand reared, cock & hens. **QUAKERS,** hand reared, sexed young prs. Reasonable prices. Orders taken for pets. Will freight. Ph. Brad **(03) 397 5278**.

GOULDIAN MUTATIONS - Australian yellows, White-breasted Normals. Will freight. Ph Gary B/H **(03) 466 2687**.

PAIR OF NANDAY CONURES, Pair of Sun Conures, 3 Pairs Tasmanian Rosellas, 1 hen Blue Ringneck, Princess Blue & Split, 2 pair of Crimson-wings Ph: **(054) 837 536**.

ALEXANDRINES - Large good quality young birds. Some hand raised. Also adult birds. Will Freight. Phone Wayne **(054) 286 626**.

PEKIN ROBINS, Eclectus, Nandays, Cock Janday. Ph **(03) 723 1866**.

HOODED PARROTS, Alexandrines, Plumheads, Crimson-wings, Princesses, Kings, Blue Bonnets, Western Rosellas, Regents &

MACAWS S/S imported surplus Scarlet, Blue & Gold. Immaculate feather, quiet. Ph A/H. Bill **(066) 449 421**

RAINBOW LORIKEETS, 1 hand reared, 2 parent reared. Ph: **(045) 721 800**

GOLDEN-CROWN CONURES, Blue & Split Crimson Rosellas. Nanday Conures. Young Plumheads. Young Quakers. Golden Song Sparrows, Chaffinches, Black breasted Quail. Ph: Alan **(057) 626 190**.

RINGNECKS Grey/green, Blue, Albino, & Green Split Blues. Sun Conures, Eclectus, Alexandrines, Plumheads & Quakers. All birds S/S, closed rung. Hand reared & parent reared available. Will freight. Ph Robert Flynn **(052) 828 254**.

SUN CONURES, hand reared S/S. Phone **(07) 962 535**.

ADULT PLUMHEADS - 2 breeding prs. Hand reared Alexandrines '93 bred S/S. 2 Cock adult Major Mitchell's. Phone Ross **(07) 962 535**.

RED-TAILED BLACK, mature male Naso, hand-raised pet, Ph **(071) 275 173**.

FOR SALE: ALMONDS & Almond products, Ph: **(08) 386 3035**.

FOR SALE: Quakers, Nanday & Conures, Plumheads. Will freight. **(054) 722 930** anytime.

WANTED

1 HEN GANG GANG Contact Carmel or Vicki-Ann on **(070) 546 447**

MATURE BLUE RINGNECKS

Left: Handraising is becoming increasingly common in aviculture.

THE GOAL

Following is a graph depicting the growth of an Eclectus chick from hatch through to weaning, showing the four distinct phases that are typical of all species.

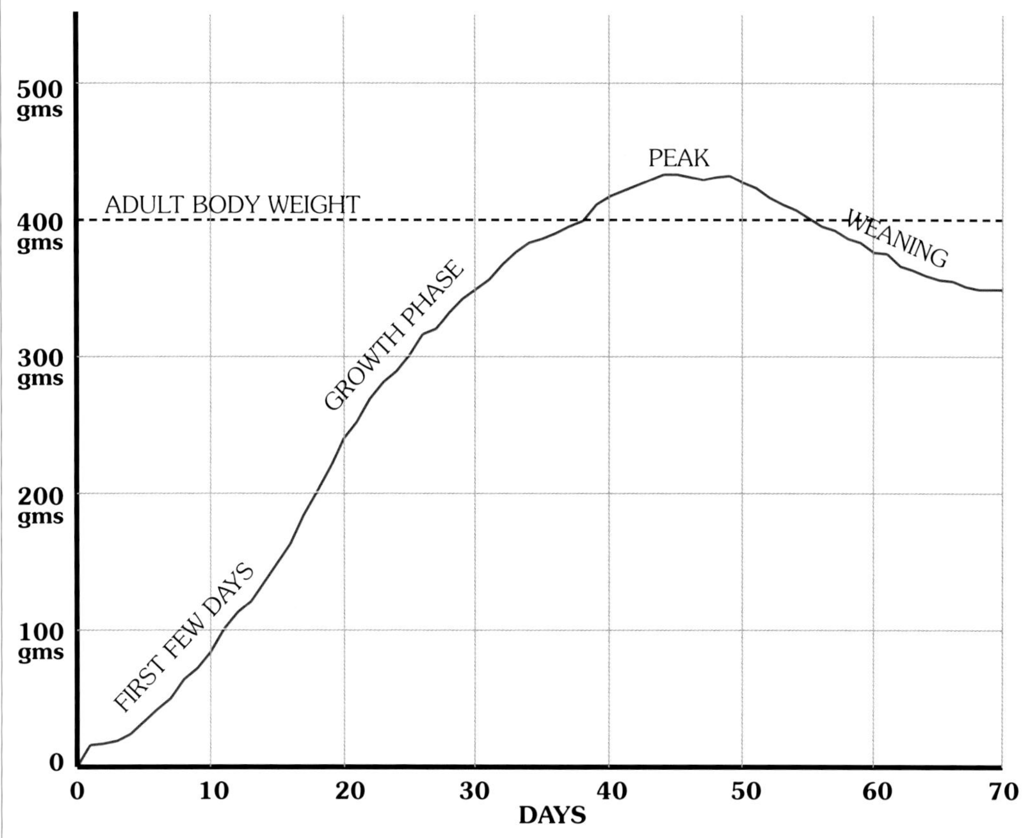

ECLECTUS E. r. polychloros
Growth Graph

1. First few days - the first few days of the handfed chick's life sees minimal weight gain and no visible changes in terms of development. However, once the chick stabilises, weight gain begins to improve as it enters the *growth* phase.
2. Growth phase - which begins approximately on day seven for most species. During this phase weight gain is substantial and the bulk of the chick's development takes place. Generally the weight gain tapers off as the chick progresses toward typical adult body weight for the species, before entering the third phase.
3. Peak phase - somewhere around adult body weight (for the species) the weight gain virtually ceases, fluctuating up and down for several days to a week, then slowly begins to drop. This is the *peak* phase which precedes weaning.
4. Weaning - a period that lasts between two weeks and five months depending on the species and ends once the chick no longer requires any handfeeding whatsoever. During this phase there is natural weight loss as the chick prepares to fly.

The above stages of growth are natural progressions, the only difference in the wild being that parent fed chicks generally gain better weights in the first few days of life. Possible reasons for this are discussed later.

PHOTOGRAPH: P. COURTNEY

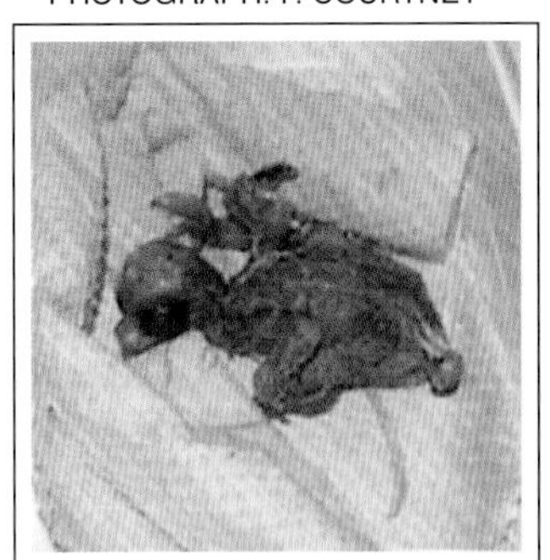

Newly hatched Eclectus

PHOTOGRAPH: CURRUMBIN SANCTUARY

Pin-feathered Eclectus

PHOTOGRAPH: CURRUMBIN SANCTUARY

Fully feathered Eclectus

Left: The goal is to produce a chick comparable in quality to that which nature would have produced.
Yellow-tailed Black Cockatoo parents with newly fledged chick in centre.

FUNDAMENTALS

Weighing

Daily weighing of the chick is essential, as weight gain is the best indicator of progress, both in terms of daily gains and also in the overall sense of where the chick peaks in relation to adult body weight for the species. Poor weight gains are an indicator of some form of problem and by regular weighing many problems will be detected long before they visually manifest themselves, making recovery easier and quicker.

Surprisingly, there are still aviculturists who do not weigh their chicks, believing that they can tell a chick's condition and progress purely by visuals eg. skin colour, eye alertness, feed response etc and while it is true that these are indicators of good health, they can be misleading. Just because a chick looks bright and the crop is emptying in a timely manner does not necessarily mean that the chick is achieving appropriate weight gain.

Weighing is particularly crucial if handraising a chick from the egg, because in the first few days of life there are little or no visual changes and only weight monitoring will tell the real story.

Right: Daily weighing of chicks is essential, as with this cockatiel chick.

No weight gain after three to four days of life or even worse, a weight loss, is cause for concern and must be dealt with. Apart from these first few days of life, there should be a recorded weight gain every 24 hours. Parrot chicks do basically all their growing between hatch and peak and it is only by gaining weight on a daily basis that the chick will reach adult body weight in such a short time (less than 60 days for the vast majority of species). No gain in a 24 hour period after the first two to three days of life is worthy of note, however not a real cause for concern. Two days of no gain strongly suggests something is not right and three days without gain indicates there is definitely a problem. Begin looking for answers on the second day of no weight gain because the quicker the problem is sourced and dealt with the better.

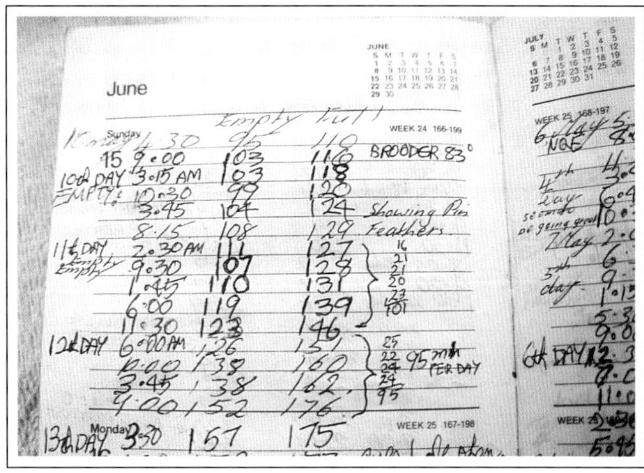

The above records, while better than no records, reveal little and are difficult to read. Accuracy, detail and clarity are necessary if records are to mean anything to yourself and others in the future.

Weighing should be performed first thing in the morning, just prior to the first feed. Generally this is the only time the crop is totally empty and an accurate weight will be recorded. Weighing chicks with varying amounts of food in the crop during the day will produce erroneous weights and therefore be of little value. Place the chick inside a container or bowl while weighing. Small chicks fall over particularly easily and may injure themselves if they roll off the scales.

Digital scales that measure in one gram increments are ideal for weighing most species, however, if the species is particularly small eg. Lovebirds, Cockatiels or Fig Parrots, then a unit that measures to one decimal point is preferable.

There are a wide range of scales suitable for the weighing of chicks and the ultimate choice will depend on two main factors, budget and the species involved.

Two ideal models used by aviculturists are Arlec™ and Soehnle™. Whatever brand is finally chosen, check before purchase that the display shows the correct weight immediately an object is placed on it. Some digital scales display several progressive readings leading up to the true weight and with a restless chick it can be very frustrating waiting for it to sit still long enough to record an accurate weight.

Typical weight gains of specific species are recorded in the tables on page 94. Suffice to say, at this point, daily weighing has many benefits and is strongly recommended. Weighing goes hand in hand with the next essential, record keeping.

Records

Whether raising one chick or 30-40 chicks, the keeping of records is an important aspect of the handraising process. The memory is at best unreliable, while records allow you to do several things:

- Accurately monitor chicks weight and development
- Experiment with brooder temperatures, feed intervals etc and thereby finetune your regime
- Monitor feed volumes to avoid underfeeding
- Go back and pinpoint possible causes of a problem
- Keep weaning times to a minimum

Following is a simple yet effective format for recording all necessary information:

Day	Weight (grams)	Gain (grams)	Feeds	Total Volume Per Day
1	18g	-	RECORD TIME, VOLUME FED in millilitres, and CROP STATUS (E = Empty, NQE = Not quite empty) eg. 2.00 pm E-10ml	12ml
2	18g	-		14ml
3	20g	2g		18ml
4	23g	3g		22ml
5	27g	4g		25ml
6	31g	4g		29ml
7	32g	1g		30ml
8	38g	6g		33ml

A quick scan down the *Gain* column will show overall progress while the *Total Volume per Day* column will help avoid underfeeding. In the *Feeds* column, record the time of feed and the volume fed at each time. This volume control is essential if crop tubing, simply because you cannot feed unknown volumes directly into the crop. By recording above each feed the status of the crop prior to feeding (E - empty, NQE - not quite empty), you will constantly be in touch with crop motility and whether or not it is time to increase the volumes per feed (eg. 2.00 pm E -10ml).

Use the opposite page to record alongside each day anything of interest or concern eg:

- Condition of chick at hatch/pull
- Formula details ie thickness, additives etc
- Brooder temperatures
- Environmental/behavioural changes
- Physical development ie eyes open, wing flapping etc
- Slow crop emptying, poor feed response etc
- First cracking of seed

The other benefit of keeping records is that you now have something valuable to contribute to aviculture. Other breeders can use such records to compare their progress in terms of weight gains, volumes, weaning times and so on, allowing them to properly assess and improve their regime. Once a format is chosen for record keeping simply run off photocopies and place in a file, compiling an information base for ongoing reference. It does not take long after each feed to record the information, which ultimately benefits the chicks, the feeder and aviculture. After all, there is still much to learn in handraising.

Hygiene

Hygiene is another crucial aspect of handraising. The greater the number of chicks in the nursery at any one time, the more important it becomes. Poor hygiene directly contributes to nursery problems in terms of illness and poor development, therefore frequent, thorough cleaning and disinfecting will keep bacterial, fungal and viral contamination to a minimum. Small chicks handfed from hatch are particularly susceptible to infection because they have not received enzymes and other gut flora from their parents, which leaves them vulnerable to all sorts of invasive and harmful organisms until they can develop their own resistance.

All feeding instruments should be cleaned thoroughly under running water after each feed then left to soak in a disinfecting agent, such as Milton™ solution or Avisafe™. Rinse in clean water just prior to feeding to remove excess solution, especially with the Milton™ solution as this may be harmful to tiny chicks. Milton™ and some other disinfectants tend to harden rubber and plastic quite quickly and so some aviculturists prefer to simply wash their crop tubes and syringes in clean water and leave them to air-dry between feeds. A compromise would be to draw in, then expel, disinfecting solution before they are stored. With any feeding instrument not soaking in solution it pays to store them in a container with a mesh lid to allow them to dry properly yet prevent contamination from wandering cockroaches and mice etc.

As with all incubation and handraising procedures, hands should be thoroughly washed before handling chicks, especially if having dealt with birds in the aviary just prior to feeding time. A convenient hand cleanser for the nursery is the liquid soap pump pack, or preferably a powerful cleansing agent such as Hibitane™. The use of disposable surgical gloves should also be considered.

Brooders, bench surfaces and the sink area need to be regularly disinfected. Have an exchange set of brooder containers on hand to replace the soiled ones every three to four days.

Hygiene in the area of diet is also very important. Any formula, whether commercial or homemade, that has been exposed to moisture should be discarded as it becomes a prime site for the growth of fungal organisms. One bad batch may result in substantial losses. Cutting corners in this area can be disastrous and any food that is questionable should be rejected. Storing and re-feeding left over formula from the previous feed, is not a healthy practice and really is not worth the risk of contamination.

A sick chick can be very frustrating to deal with, not to mention the anxiety over recovery, needing a separate mix, regular medication and individual brooding. Smooth, professional handraising is about eliminating risks and attention to detail in the area of hygiene will contribute greatly to a problem free nursery.

BROODERS AND BROODING

Brooders

A brooder is simply an enclosure/cabinet fitted with a heat source and optional fan, in which chicks are brooded until they develop feathers and can regulate their own body heat. Depending on the age and number of chicks that require brooding, a cheap, basic unit will suffice or perhaps several more expensive and elaborate units will be required. Certainly more than one unit becomes necessary where handraising is over an extended period with chicks of varying ages. The reason for this is that as chicks grow older they need lower temperatures and various species require different temperatures at the same age.

Many poultry/avian dealerships now sell brooder kits consisting of a thermostatically controlled heater unit and small cabinet type fan for distributing the warm air. Providing you have a few basic tools, the cabinet can be easily made. If constructing your own cabinet use a material that has a smooth surface and is easily

Right: You can buy a brooder kit, electric fan and heater element and make your own cabinet.

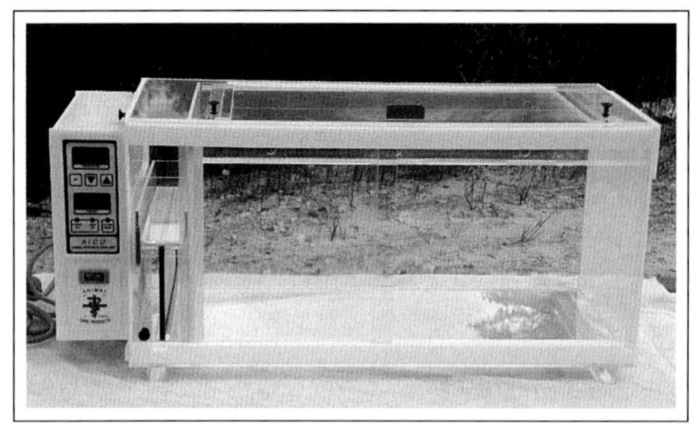

Left: A widely used fan forced brooder, the Lyons™ Intensive Care unit.

disinfected eg. perspex, plastic or laminated timber. Porous surfaces are difficult to effectively clean and therefore unsuitable.

Fan forced brooders are preferred over still air models when dealing with small chicks requiring a more uniform and stable temperature. Chicks up to ten days old are quite susceptible to fluctuating and marginal temperatures and this can be a problem in some still air units, as already discussed in the section, *Choosing an Incubator/Hatcher*.

The other advantage of the fan forced unit is that when a chick is placed back in the unit after feeding, it is warmed quicker. When removing a chick from a brooder for feeding and with young chicks that will be from eight to ten times a day, much of the warm air is lost. When placing that chick back into the still air unit it can often be observed shivering for a few minutes until the air around it warms up. This regular chilling can cause slow crop movement. Be aware that still air brooders which supply heat from one side only, can have significant variations in temperature around the unit and will also be markedly affected by external temperatures. Fan forced brooders that are popular and suitable for smaller chicks include the Lyons™ Intensive Care Brooder, the WAPE™ Parrot Brooder, the AB™ Newlife Brooder and the Brinsea™ Octagon 20 Parrot Rearing Module.

Above: As chicks develop, heat accuracy becomes less important and inexpensive hospital boxes will suffice as brooders.
Below: The fan forced WAPE™ Parrot Brooder offers all-round viewing.

Brooding

As mentioned, brooder temperatures are critical in the first week or so of a chick's life and a one to two degree adjustment can often make all the difference in weight gains and crop motility. This need for accuracy lessens as the chick develops, to the point where at late pin-feather stage a heat source may not even be necessary.

Regardless of age, the rule for the brooding of chicks is this -

visual observation and instrument readings should be used together to find optimum brooding temperature.

Impact of brooder temperatures on weight gains and crop motility is direct. A chick that is too cold will devote more of its energy into staying warm than putting on weight and consequently digestion will slow. A chick that is too warm will be restless and move around rather than sleep. Activity will chew up potential weight gain and stress will slow the crop. The best temperature setting is one where the thermometer is somewhere close to recommended temperature and where the chick is comfortably sleeping most of the time. Commonsense prevails here, a chick shivering or huddling in a ball is too cold and a panting, restless chick is too hot.

Above: Homemade brooders successfully used by Perth Zoo, WA.

Below: Weaning cage fitted with a Crompton™ Fire Lamp.

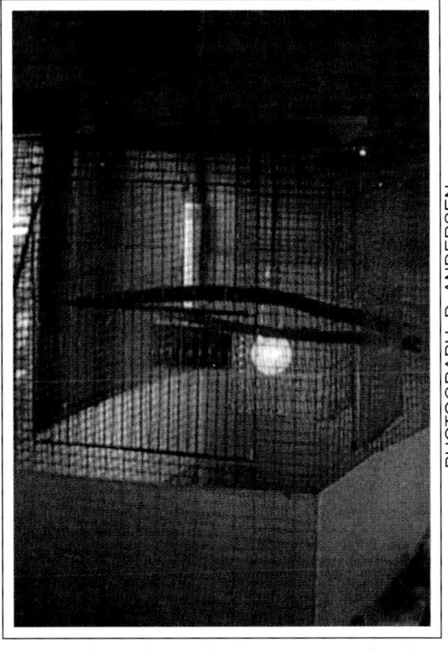

PHOTOGRAPH: D. ANDERSEN

Following are temperature guidelines for the brooding of chicks:

Newly hatched	36.6°C
5-12 days	35° - 31.6°C
12 days - pin-feather	31° - 28°C
Once feathers begin to cover most of the body	26.5°C.

An example of why the above temperatures are purely guidelines, is that a White-tailed Black Cockatoo chick with heavy down is able to tolerate 30°C at nine days of age, while a naked Major Mitchell's Cockatoo chick of the same age would still require 34°C and for this reason, the visual observation of the chick is important.

It is suggested that chicks be brooded individually for the first few days until stabilised and gaining weight. This allows monitoring of the individual's faeces and behaviour, important indicators of wellbeing in the early stages. While brooding chicks singularly, place a small furry toy about the size of the chick in with it. This satisfies the chick in terms of a companion/mother and the chick will often brood up against it. Nobody wants to grow up alone!

Where chicks are brooded together, recommended after the first week or so, the communal behaviour is as good a guide as any in finding optimum temperature and particularly useful in the absence of a thermometer. A tightly packed group where all the chicks are pressing into the centre of the group indicates they are too cool and where the chicks have all moved to the extremity of the container, they may be too warm. Ideally the chicks should form a loose cluster. Chicks brooded together, besides being happier, will also tolerate a slightly lower temperature and more so if pulled from the nest. Parent reared chicks can tolerate lower temperatures compared to their nursery counterparts of the same age, having become accustomed to temperature

Left: All your brooding equipment - various sized plastic containers and bins, tissues, sawdust.

fluctuations from an early age, as the hen has been on and off the nest. Take this into account when pulling a chick from the nest. To put it into a brooder set on 'recommended' temperature, may actually be too warm and upset the chick.

Try to avoid brooding in an overly bright environment. Consider the natural nest situation in the wild, where most nests are in a hollow branch/trunk and are either dark or semi-light. If the brooder is one that uses light bulbs for the heat source, then this light needs to be diffused or reduced in some way to produce chicks that sleep better and as a result grow slightly better. A very fine mesh around the globe or a cardboard box over the brooder container will simulate the natural nest. However, if placing a box over the chicks, be sure that it has plenty of holes punched in it for ventilation. Further, be aware that if it is a still air brooder, the temperature inside the boxed area will differ from the general brooder temperature, therefore, it is advised to have a thermometer inside the boxed area.

Humidity is not as important in brooding as in incubation, however a chick will suffer if left in an excessively wet or dry environment. Placing a small bowl or dish of water with a surface area similar to that of a small margarine container will provide sufficient humidity to maintain a healthy chick. This is necessary in a fan forced unit to prevent dehydration. Avoid excessive humidity in the brooder as this creates the perfect setting for harmful organisms to flourish. Mention should be made regarding brooding Eclectus. Do not be concerned if the skin is noticed flaking off the chick at around 8-14 days of age. This is a natural phenomenon in developing Eclectus chicks and has little to do with humidity levels.

Above and below: To virtually eliminate the possibility of woodchips being swallowed, use wire mats to pack down the chips.

Containers

The vast majority of breeders use plastic containers of one description or another for brooding their chicks. They are cheap, easily cleaned and long lasting. Simply progress from small containers for the individual chick up to ice-cream containers or buckets as the chicks are clutched and grow larger. For the larger cockatoos the milk crate sized bins available from most toy shops are excellent.

Above: Choose a suitable grade of non-toxic sawdust to avoid problems.

A relatively small container for the first week or so of a chick's life can be beneficial for several reasons. It will help the chick to sleep upright, reducing the risk of aspiration from sleeping on a crop full of fluid and possible forcing of formula back up the oesophagus. It also helps prevent the chick's legs from continually sliding out from under it and possibly developing splayed legs and thirdly, a chick is generally more settled and secure in a small container surrounded by tissues or a soft toy.

Once chicks have reached late pin-feather stage there is no need to supply heat, particularly if two or more chicks are together. To keep them warm at night, especially in some areas of Australia which can be quite cold, simply place a cardboard or thin ply lid partly over the bucket or bin to keep most of the body heat in. In a plastic container, where there is no real air flow, the combination of body heat, droppings and plastic produce extremely high humidity to the point where condensation actually runs down the sides of the container. This is extremely uncomfortable for the chick and can be avoided by leaving the lid partly to one side to allow the container to 'breathe' and by punching a few ventilation holes in the container itself. These lids are also handy for writing information on regarding the chicks inside.

If brooder space is tight and chicks need to be evacuated from a brooder early to make room for the next batch coming along, there are several options. A foot-warmer pad can be

Above: With single chicks, brood with a furry toy for companionship.
Right: Once pin-feathered, chicks brood well on a wire base where faeces can fall through.

Left: *Large plastic bins suitable for brooding cockatoos.*
Below: The chick second from right has locked itself onto the rim of the container.

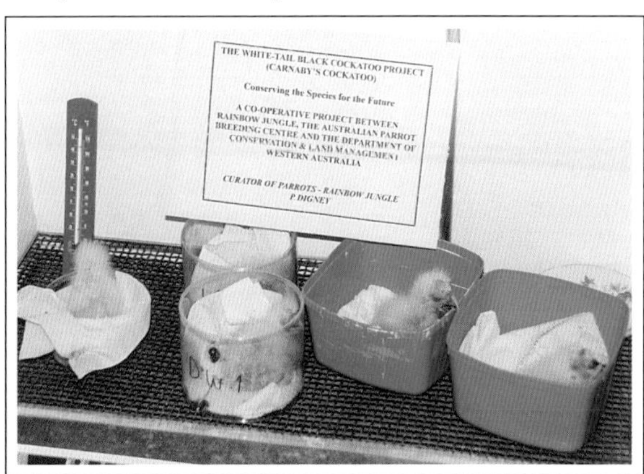

hung down the side of a crate or cardboard box, the chicks having the advantage of moving away or closer, depending on their heat requirements.

Alternatively, there are a variety of economical clip-on lamps that can be attached to the side of the brooder container, facing down. Elevate the light so that the chicks cannot burn themselves on the globe. A 60-80 watt light bulb will supply sufficient heat to keep chicks warm, however use bulbs that are pearled or coloured to reduce the glare that may otherwise stress the chicks. Some lamps are not designed to take a light bulb stronger than 60 watt and this needs to be checked at purchase time. Heat lamps that give off infra-red warmth are preferred over light bulbs because very little light glare is given off. The Crompton™ Fire Lamp is a suggested alternative.

Some chicks develop a habit of locking their beaks onto the lip of their brooding container and if this behaviour develops it may be necessary to place the chick into a container where the lip is out of reach. This scenario is discussed in more detail in *Troubleshooting - Underfeeding*.

Bedding

For the first few days of life the small chick is best brooded on tissues which are replaced every feed. Droppings will vary greatly in colour and descriptions depending on diet. When changing the tissues every feed, any unusual features or changes will be detected immediately. Use several layers of tissue at a time, as a single layer soon breaks up with the moisture from the droppings and the chick will begin slipping on the undersurface.

Once the chick enters the growth phase, sawdust is by far the most popular bedding used. It is cheap, easily obtained and an excellent absorber of fluid. The one problem with this is that from pin-feather stage many

Left: Handfed White-tailed Black and Major Mitchell's Cockatoo chicks.

Right: Blue and Gold Macaw chicks on an elevated wire base to prevent ingestion of woodchips.

chicks, particularly cockatoos, begin picking at the base material which can result in a crop full of compacted woodchips. Even the swallowing of one large or sharp splinter may cause crop damage and slowing. There are two ways to largely minimise this problem. Firstly, choose the type of sawdust carefully. The larger woodchip is less absorbent and more likely to be picked up, chewed, then swallowed, while extremely fine sawdust which is powder-like can produce airborne dust that may be inhaled and irritate the chick's respiratory tract. A little time spent visiting garden nurseries and timber merchants will assist you in sourcing a toxic free sawdust that is not too heavy, not too light, but the desired grade.

The other way of preventing the swallowing of woodchips is to cut a mat of wire which fits neatly inside the brooder container on top of the sawdust. By using 12mm x 12mm, 18 gauge wire it will act as a cover and pack the sawdust down. The smaller the aperture of the wire the better as it prevents the chick picking at the base while still absorbing the dampness from the droppings. Have several pieces spare and every few days drop in a new one and leave the soiled one soaking in disinfectant.

Basically, the above materials and methods work so well that there has been little experimentation with other types of bedding, however crushed popcorn, untreated pine bark, vegetable based kitty litter and even maize/wheat have also proved successful. Once the chicks begin to feather up, a simple wire floor that is elevated to allow faeces to drop through onto a tray or newspaper works well providing the aperture of the wire is small enough to prevent feet and toes being caught.

FORMULAS

Congratulations to all those aviculturists who endured the 'rolled oats, weetbix and honey' formula years. While in essence they worked, there were attendant problems with such basic diets for a growing chick. Fortunately there have been significant advances in the area of handraising formulas for parrots in recent years and with the advent of 'add water and stir' mixes, handfeeding has never been easier.

Such manufactured diets are balanced in terms of protein, fat, energy and vitamin/mineral content, therefore supplying all the necessary nutrients for mainstream parrot species, eliminating many of the growth and crop problems of yesteryear. Some of the more traditional homemade formulas still persist however, and in fact, compare quite favourably in terms of the end result. Then of course there are individual species in Australian aviculture whose dietary requirements are met by neither of the above and need a tailor made formula eg. Fig Parrots, Glossy Black Cockatoos.

If handraising is a new experience, do not expect to simply mix up a formula and achieve outstanding results with the first chick. Having made a choice of formula and

Left: Many excellent commercial formulas are now available to the aviculturist.

knowing that it has proved successful with others, stick with it. Spend time experimenting so that you become intimate with the mixing temperatures, dilution ratios and additives that produce the best results. This will take time and accurate records, but remember that it is all about developing a long term, hassle free regime that works well. While time is spent changing whimsically from formula to formula no substantial progress will be made and what should be an enjoyable hobby may quickly turn frustrating.

An intriguing occurrence in many nurseries is the knee-jerk reaction to problems with chicks - blame the formula. So often the formula, particularly the commercial ones, become the punching bag immediately things go wrong, when more often than not the problem lies elsewhere. This is important to appreciate because while the focus is on the formula, the real cause of the problem will remain undetected. As will be discussed in the section on *Troubleshooting*, there are a whole host of reasons why a crop is slow, why gains are poor etc and more often than not it turns out to be an operator-related problem. So if something is not right in the nursery, sure, consider the formula as a possible cause however, do not dwell on it alone.

Commercial Formulas

There are now several popular commercial formulas available in Australia, some of these being Lakes™, Vetafarm™, Roudybush™, Wombaroo™, Pretty Bird™ and Loristart™, amongst other less widespread formulas. Some of these proprietary brands produce species specific handrearing formulas for lories, lorikeets, pigeons etc.

The vast majority of Australian parrots/cockatoos do well on any of these formulas. The chief advantages are:

- Ease of preparation
- Most formulas pass through a crop tube easily
- Balanced nutritional levels

Follow the manufacturers instructions when mixing commercial formulas, particularly Lakes™, which has quite specific guidelines to follow. Never mix a formula of any sort with overly hot or boiling water because not only will it change the true consistency, turning it gluggy, it will also tend to diminish the value of the all-important nutrients. On the other hand, mix Lakes™ too cool and it tends to settle out on the bottom quite quickly.

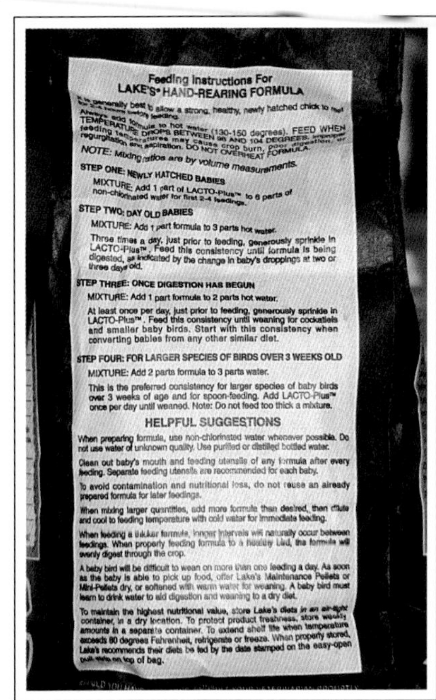

Right: Instructions for mixing of Lakes™ Handrearing Formula. It is important to always follow the manufacturer's instructions.

All the major commercial formulas have a smooth consistency once mixed, however both Lakes™ and Wombaroo™ have a fine grit element that needs to be sieved out if using a very fine crop needle or small syringe. All can be spoon fed, however Lakes™ when mixed at recommended ratios for various ages tends to be quite a bit thinner than other formulas so is more suited to syringe and crop tubing.

All supply balanced levels of fat, fibre and protein to the growing chick.

Brand	Protein	Fat	Fibre
Lakes	17%	5%	4.5%
Vetafarm	22%	14%	10%
Roudybush	21%	7%	5.5%
Wombaroo	22%	8%	5%
Pretty Bird	19%	8, 12 or 15%	2%

Homemade Formulas

The only discernible advantage to homemade formulas over commercial ones is pricing, being cheaper to prepare yourself. Perhaps this advantage is traded off, however, when one takes into account the running around and time to purchase and mix the ingredients. Still, that remains an individual choice and while more and more breeders are opting for commercial mixes, some of the old recipes still persist and produce fine chicks. A problem with the majority of homemade mixes is that unless finely blended, which takes some effort, they are generally too coarse to pass through crop tubes and syringes. Following are two recipes for those who choose to prepare their own.

Above: Most commonly used formula additives - peanut butter, apple sauce, steak and vegetables, beef and vegetables.

RECIPE 1

1 cup raw wheat germ
1 cup high protein dog biscuits (ground fine)
1 cup hulled sunflower seeds (ground fine)
1 cup corn (maize) meal
1 cup millet meal
2 cups high protein Farex™ baby cereal.

or

RECIPE 2

1 cup ground sunflower seeds
1 cup ground almonds
4 Granita™ biscuits (blended)
1 packet high protein Farex™ baby cereal
1 tablespoon glucose powder
1 cup corn (maize) meal

Where necessary place the above ingredients in a blender to grind to a fine consistency.

Right: Never mix formula with extremely hot water. Not only does it turn gluggy, it destroys many of the nutrients.

This task is made easier with the sunflower kernels if they are frozen first. Once all the ingredients have been blended, mix the recommended quantities together then store in a sealable plastic container. If larger quantities have been mixed up then it is suggested most be frozen and a week's supply is taken out as necessary. Whether frozen or shelf stored, the key here is to make sure the container is airtight to maintain freshness and quality.

Some aviculturists who have changed from their traditional homemade formula to a commercial mix, experienced early problems and have reverted back to the homemade mix. Again, it is all about becoming familiar with the particular traits and performance of a particular formula, which takes time. Remember, others are using commercial formulas successfully! Also note that most homemade formulas have a coarser consistency than smooth commercial mixes and will naturally be digested slower, so do not be alarmed if volumes fed per 24 hours do not match those of the commercial mix.

Additives

Additives can enhance the performance of a particular formula depending on what and how much is added. Two additives widely used are apple sauce and beef/steak and vegetable. Both are Heinz™ baby food products and can be purchased at most supermarkets in small jars. When buying the beef and vegetable or steak and vegetable, check that it is the mix for infants four months and younger, as the mix for older infants contains solid lumps of vegetable that make it impossible to crop tube or syringe feed. It is also suggested that if the food is in tins that it be transferred to a sealable glass container upon opening to reduce the possibility of contamination that has been linked to such tins.

Apple sauce contains the enzyme pectin, which has an astringent or drawing effect on the muscle wall of the crop. This pulling-in, as it were, tends to force food into the digestive tract that little bit quicker. The faster the crop empties the better for the chick in terms of health and weight gains. The addition of beef/steak and vegetable is simply to supply meat protein and a direct vegetable element. Beef and vegetable (not steak and vegetable) has an interesting effect on Lakes™ Handrearing Formula when added. If the formula is left to sit for a couple of minutes after mixing, an enzyme in the beef and vegetable breaks down the formula and it turns noticeably thinner. This has allowed some breeders to feed Lakes™ with a higher solids content than suggested by the manufacturers and produce excellent gains in older chicks. However, adding too much of any of the above will reduce and dilute the essential value of the formula. The base formula should always form the bulk of the volume. A suggested proportion for these additives is approximately one rounded teaspoon of each per 75ml of formula.

Adding peanut butter to the formula, as many aviculturists do, increases the fat content. There is no doubt that the addition of peanut butter does produce better weight gains and while it has been claimed that all it is doing is adding body fat to the chick, it needs to be noted that peanuts contain 25% protein. There can be no doubt that this contributes directly to greater tissue and bone growth. Some species require more fat than normal in their diet and in fact need the addition of peanut butter eg: Black Cockatoos, Macaws, Glossy Black and Palm Cockatoos. Digestion does slow down a little with the addition of peanut butter and it is not recommended for chicks under seven days old because in those early days the goal is to get the crop moving as fast as possible in preparation for the growth phase.

The practice of combining two or more commercial formulas for anticipated better results appears to be largely futile and may, in fact, be detrimental in some cases. Depending on which formulas are combined and in what proportions, something beneficial in one formula may be diluted to the point of little value to the growing chick. Having said this, some breeders firmly believe that they have achieved better results and gains by combining formulas. If it works, then stick with it.

Calcium

All of the commercial formulas claim to contain sufficient calcium for proper development of the chick, however, homemade formulas may be in need of supplementation. Some aviculturists simply crush cuttlefish and add a pinch to the formula each day. Other supplements are Calcium Sandoz™ syrup and Calcivet™. All are excellent sources of calcium and its addition to the formula is recommended. However, be aware that over supplementation of calcium can actually stop calcium absorption in the intestine, so wherever possible follow the manufacturers dosage recommendations or consult an avian veterinarian.

Above: Three excellent calcium supplements for the growing chick.
Below: Probotic™, containing eight strains of friendly bacteria, is recommended for young chicks and birds suffering slow moving crops or sickness. Spark™ on the right, an excellent rehydration supplement.

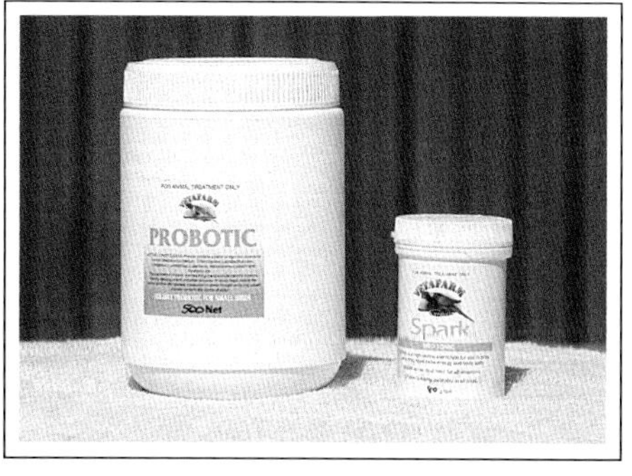

Multivitamins

The addition of a drop or two of multivitamins to the formula appears to be largely unnecessary to produce a quality chick unless the diet is a very basic one. It is suggested that very small chicks are not fed multivitamins because the concentrated nature of the additives can be damaging to the liver. If choosing to add multivitamins, use sparingly and not until the growth phase.

Chlorophyll

Spiralina, available in powder form from the health shop is one of the most complete nutritional sources known to man and is increasingly being added to formulas. It contains, amongst other important elements, chlorophyll and a pinch of this powder placed in a cup of formula is an excellent addition. Processing celery or blanched silverbeet through a juicer or blender also produces liquid chlorophyll and a little added once or twice a day to the formula is a recommended alternative.

Natural Flora

The parent fed chick is constantly receiving friendly bacteria and natural gut flora from the parent via regurgitated food which contributes to the much better start that parent fed chicks display in the first few days. To compensate for this in the nursery, it is common practice to add some sort of live culture to the formula such as acidophillus lactobacillus powder or Probotic™, an Australian product containing eight different strains of friendly bacteria. It is strongly recommended that a pinch be added to every feed for the first few days of life and to any bird suffering slow moving crop or sickness, including adults.

There is some debate as to whether these organisms in fact colonise the crop and digestive system of parrots or whether they actually pass through. However, if they are present continuously in the small chick via every feed then they may well go a long way to preventing harmful organisms taking up residence until the chick develops its own natural resistance. More research is needed in this area, however, the addition of

Probotic™ does appear to have a positive effect on crop motility and can certainly do no harm.

Water

Opinions vary greatly on how far to take the 'purity' line with the water used to mix the formula each feed and at the end of the day it becomes an individual choice. However, at least for the first week of the chick's life, it pays to take a few extra precautions in terms of the water quality. Whether it be tap, spring or distilled water, it should be boiled first when used for handraising. Store in the fridge in a sealed container. Water from a rainwater tank, although perceived as better water, is actually more likely to contain harmful organisms than treated mains water and is best avoided unless well boiled first.

FEEDING INSTRUMENTS

Consistency of formula, number of chicks being raised, time available and whether the intention is to raise pets or breeders are all factors that influence the final choice of feeding instrument. Whatever instrument is chosen, there will be times during handraising where a change of feed method becomes necessary. Ideally, the handraiser should be proficient with the operation of all three options, the spoon, syringe and crop tube.

Advantages and disadvantages are:

	Advantages	**Disadvantages**
Spoon	Produces quieter birds Accommodates any formula Most pleasurable Simulates natural feeding Easiest method	Possible imprinting Messy Slow No chick control No volume control
Syringe	Cleaner than spoon Quicker than spoon Produces quieter bird Volume control	While quicker than spoon still slow No chick control Danger of killing chick through aspiration Will not accommodate lumpy formulas
Crop Tube	Quickest method Cleanest method Total volume control Avoids imprinting Total chick control	Will not accommodate lumpy formulas Difficult to use on some weaning birds Bird not quite as tame unless extra time is spent at feeding

Spoon

The spoon is undoubtedly the most widely used instrument in the handraising arena. Simply, it is a

Above: Spoon feeding. Simply bend up the sides of a teaspoon or tablespoon.
Left: Spoon feeding can be messy. Clean the chick's face with tissues after each feed.

teaspoon or tablespoon with bent up sides to allow the formula to flow off the end into the chick's beak. Spoon feeding is a very enjoyable and interactive experience as the chick pumps away, emitting squawks of delight and flapping its wings. It is also the method that most closely simulates the natural feeding of the chick by the parents. One foreign species that really is fun to spoon feed is the Quaker or Monk Parrot. It is a particularly playful and vocal species and literally bounces around the bench at feed time. Should the desired result be an exceptionally tame family pet, then the spoon is the best way to build a feeder-bird relationship due to the time spent with the chick. It also, along with the syringe, is the easiest method of feeding and can be performed by any member of the family.

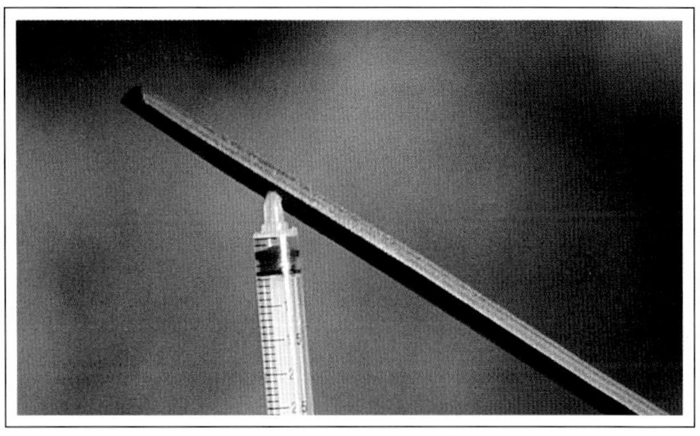

Above: If syringe feeding small chicks it will help to file the tip of the syringe so that it fits inside the beak.

Below: The Bovivet™ Plexi range of custom syringes ranging from 5ml to 50ml.

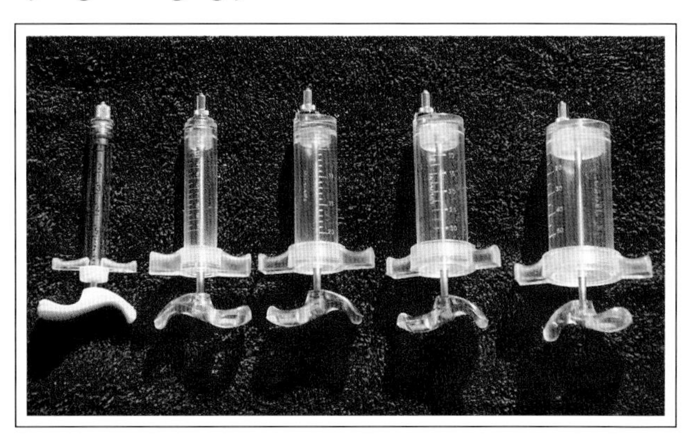

To spoon/syringe feed a chick simply hold the head gently with thumb and forefinger either side of the beak and at the base where top and bottom beak meet, while gently wrapping the other fingers behind the head for support. Once a little food is dribbled onto the tongue, the chick will begin a feed response known as pumping, at the same time swallowing the formula. Providing the chick is pumping, keep the formula flowing off the spoon but not so fast that it backs up the mouth and spills out the beak. After each spoonful allow the chick to catch its breath before repeating the procedure until the crop is full. Spooning can be messy, particularly with the larger species that have a vigorous feed response. Most spoon fed chicks will need a clean up around the beak with a tissue after each feed. Not that this is a problem, however, if left unchecked will result in a rather scruffy looking chick, food matted around its face. Spoon feeding is also a slow process and for this reason not a lot of the larger handraising nurseries use this instrument, preferring to syringe or crop tube feed.

If a chick is sick, pulled late in development or weaning, it may be difficult, even impossible to feed, rendering the spoon rather useless. Most chicks in these situations will resume feeding after experiencing a little hunger. However, the sick chick often will not and only crop tube feeding will save such a bird.

Depending on the approach you ultimately decide to take in handraising, volume control may or may not be important. However, it is highly recommended (the benefits of such are explained shortly) and there are two ways to determine the amount of each feed when spooning. Knowing that basically one millilitre of formula = 1 gram, either weigh the chick or the feed container before and after each feed and record the difference in millilitres fed.

A bird, that when independent, shows more affection and attention to its keeper than others of the species is termed to be *imprinted* and this can be the result of spoon feeding. If a family pet was the goal then this is excellent however, if the intention was to produce a breeder then that breeding ability is often, although not always, compromised, particularly with cocks. The other side of the coin is that very tame hens often make better sitters. To avoid an overly-tame or imprinted chick if spooning, spend only as much time with the chick as necessary at each feed. Basically the less it sees you the less it will become attached to you.

Syringe

Syringe feeding is a compromise between the spoon and the crop tube and in many senses is the best of both worlds. It still produces a quiet bird and is quicker and cleaner than spoon feeding. As long as the chick is pumping, the formula can be pushed into the mouth, meaning more can be fed in a shorter time. Further time is saved if a large syringe that can suck up enough formula for three to four chicks at a time is used, while one is catching its breath another one can be fed.

Volume control is one of the real benefits of the syringe. The 1ml syringe is an excellent instrument for feeding newly hatched chicks because the increments provide volume knowledge to 0.05ml and as the chick grows so does the syringe size. A potential danger when using the syringe is that to continue pushing food into the chick's mouth after it stops pumping, may cause aspiration and death.

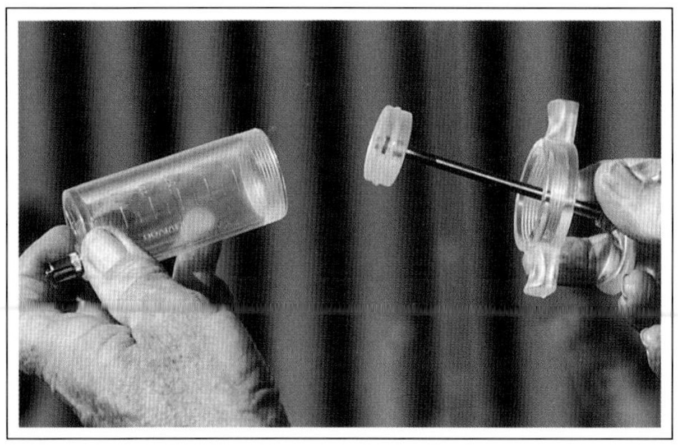

Above: These syringes are long lasting with replaceable washers and millilitre increments embossed on the cylinder.
Below: Spoon attachments for the Bovivet™ Plexi range. These produce the advantages of both spoon and syringe.

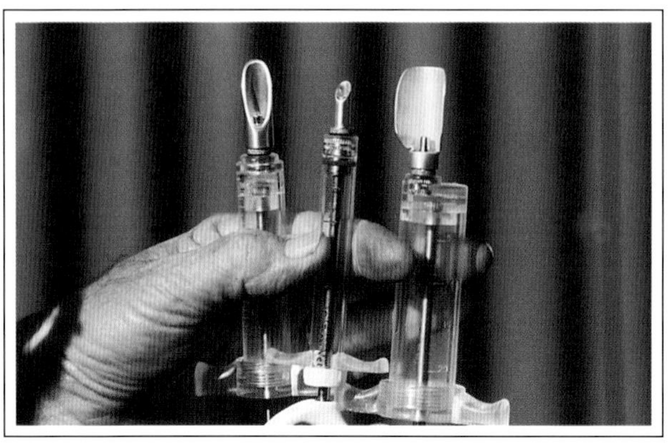

Crop Tube

Crop tube feeding of chicks, while practised in many forms for decades has only recently begun to gain wider acceptance. It is versatile, extremely quick and once confident in its use, is a very easy procedure. However, it does differ markedly in application from both the spoon and syringe. With this in mind, along with the fact that there is a lot of apprehension over its use, it is considered appropriate to devote quite some text to its role and practical use in handraising.

The term crop tube refers to a flexible tube attached to the tip of a syringe which is long enough to expel formula directly into the crop. Ideal feed tubes are the Folley™ or Nelaton™ catheter tubes available in various diameters from most surgical supplies and some chemists. The recommended tube sizes for very small species is 10-12 gauge, species up to the size of cockatiels is 14 gauge and for larger species is 16 gauge. Over time and with use

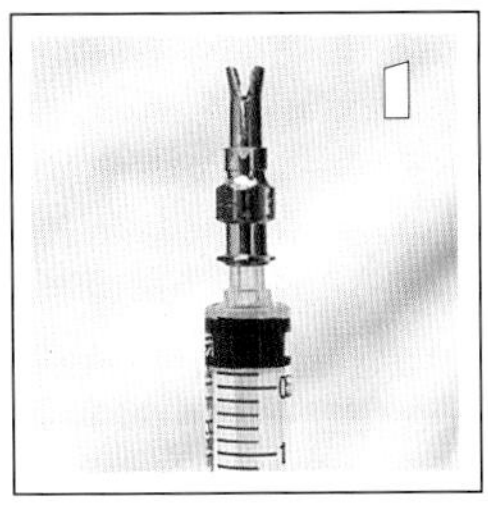

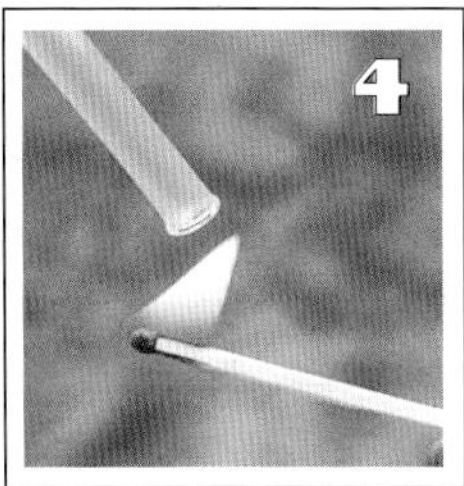

The feed tube must be attached securely. Firstly, flare the tip (1), or if the fitting is brass, file the neck down, (2) then clamp the tube with a small cable tie (3). Finally, round the insertion end off with a flame (4).

in hot water this soft, flexible tube begins to harden and therefore needs to be replaced every week or so.

Excellent syringes for crop tubing are the custom Bovivet™ Plexi range. They have interchangeable stainless steel tips, replaceable nylon washers, embossed measurements on the cylinder for volume control and range in size from 5ml to 50ml. Australian made too! The cheap disposable syringes from the chemist are not recommended in on-going crop tubing for several reasons. The volume measurements quickly rub off, the washers deteriorate after just three to four feeds and the tips are not really suitable for secure attaching of the tube.

This clamping of the flexible tube to the nipple is important because no matter how tightly it is pushed on to the tip, as it hardens it will easily slip off, particularly if the chick is pumping. The result will be a tube in the crop, which will be at least, stressful to the bird and if left there for too long may move down into the proventriculus, resulting in death. It has happened before! One way to firmly attach the tube to the tip is to slightly flare the end of the nipple, then using a small cable tie, clamp the tube around the neck. Replacement is simple, just cut the cable tie off, remove the old tube, slide the new one on and re-clamp. An ideal tube length for smaller species is 40mm and for larger species 70mm. Any longer makes it difficult to control the tip when entering the beak. Round the insertion end off with a quick flame, however, avoid breathing the fumes as they are highly toxic.

There really is no role for the fixed solid tube in crop feeding, except where the bird is an adult that must be force fed and is strongly resisting entry of the tube. Considering that with many species the tube will travel up and down the oesophagus as many as 200 times before weaned, a solid

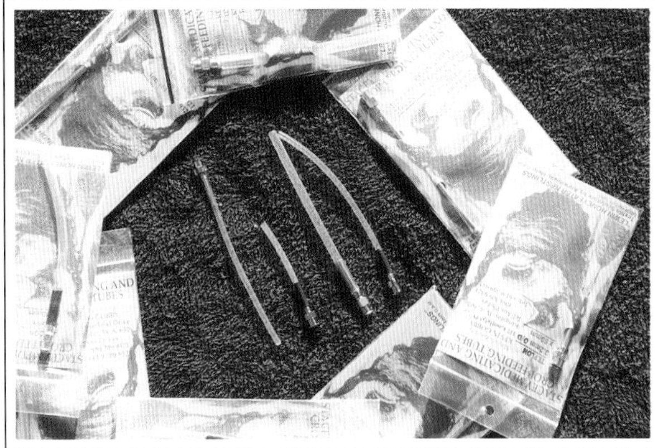

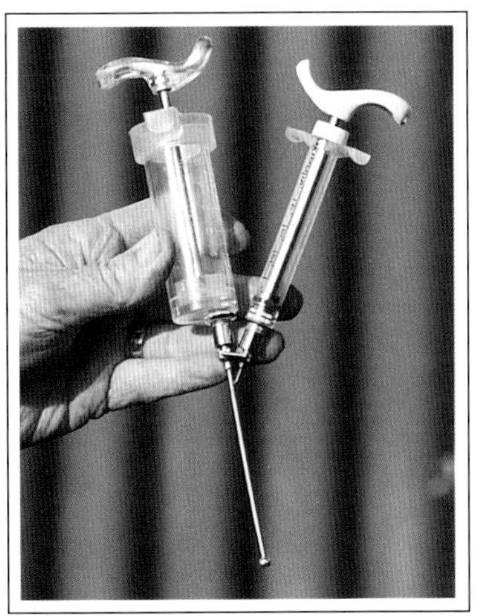

Above: Some of the tube options available in the Bovivet™ Plexi range.
Left: The crop rinsing tool. This requires two people to operate, however, rinsing the crop becomes very quick and simple.

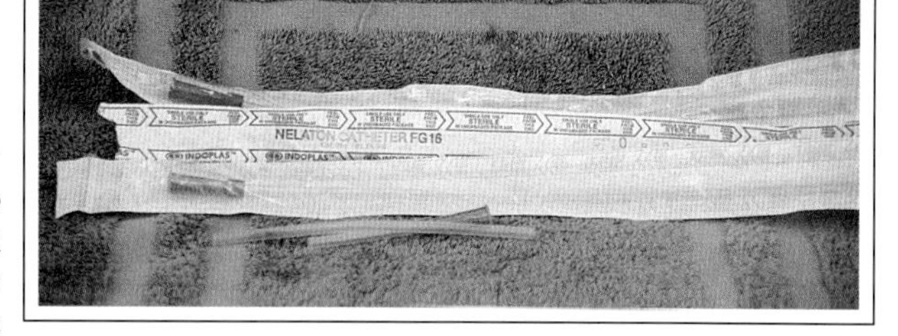

Right: Recommended tube for crop tubing, the Folley™ or Nelaton™ catheter tube is available in various gauges.

tube is much more likely to cause an irritation or injury than a soft flexible one. A flexible feeding tube will accommodate entry from an incorrect angle and will also allow for movement of the chick while feeding, a fixed tube will do neither.

Before leaping head first into crop tubing, the layout of the bird's mouth and the mechanics of insertion need to be understood. As the diagram shows there are only two ways to go with the tube, either down the oesophagus canal and into the crop (the right way), or down the trachea/windpipe (the wrong way!). The oesophagus canal actually travels down the bird's right-hand side of the neck and entry is via the cavity comprising the rear of the mouth. The entry to the trachea/windpipe is termed the glottis and is located at the very back of the tongue and at the front of the lower mouth cavity. Next time a fair sized parrot dies in your collection, open its beak fully and pull its tongue forward firmly, the glottis will be obvious at the very base of the tongue. This physical examination of the bird is important in adding to your confidence and is a bit like reading the road map before going into strange territory. You do not end up in the wrong place!

Once the location of the windpipe is understood, simply travel to the rear of the mouth before going down the bird's right-hand side (your left). Initial entry into the beak is from the bird's left, so that the tube is slightly angled to the bird's right as it proceeds down the oesophagus. Once the chick is trained to take the tube (generally within two days of tube introduction), the moment the tube enters the mouth, the chick will fully open the oesophagus and literally swallow the tube. In this situation and whenever there is a feed response, the entrance to the windpipe is automatically closed off, making it impossible to go the wrong way.

If the tube is being introduced to a chick that has previously been spoon or syringe fed it really is an easy operation. Using the fingers to hold the chick's head as you would for the spoon, initiate a feed

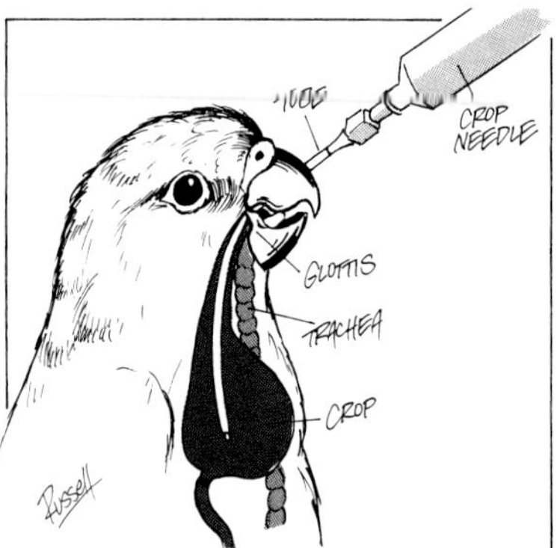

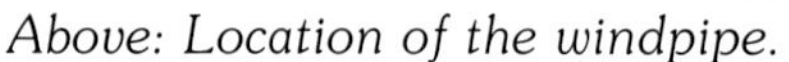

Above: Location of the windpipe.

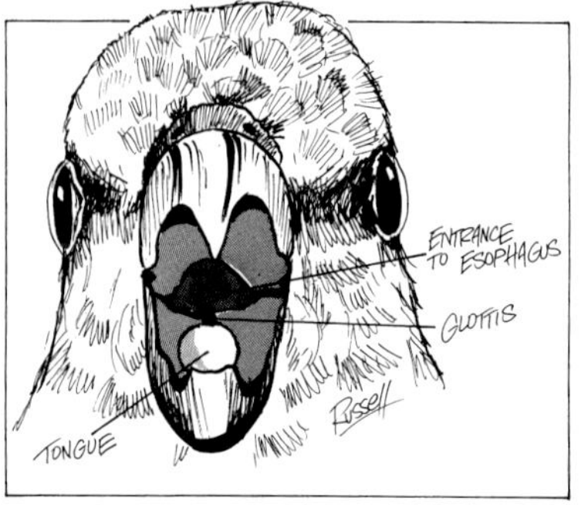

SKETCHES BY RUSSELL KINGSTON

response, then lower the tube into the crop and expel the formula. As in the diagram, there is no need to go all the way to the bottom of the crop, in fact this should be avoided. To 'bottom out' will produce a natural regurgitation reflex and will also possibly cause injury. However, the tube needs to be far enough into the crop (approximately one third) so that as the crop is filled, the formula does not back up the oesophagus and into the mouth. Should the chick be one pulled from the nest, it may take a day or so to accept the crop tube but will soon happily sit there while the crop is filled. With a stubborn chick, in the early stages of introduction it does help to spoon feed a little at first. This relaxes the chick and while it is still thinking about its next mouthful, slip the tube in. The beauty of the flexible tube is that providing you are generally heading in the right direction it will find its own way into the crop. Lubrication helps the tube slide down that little bit easier. Dipping the tube in formula will suffice, however, other options are cooking oil or Calcium Sandoz™ syrup. Dipping of the tube tip into a small cap of either will make introduction that little bit easier.

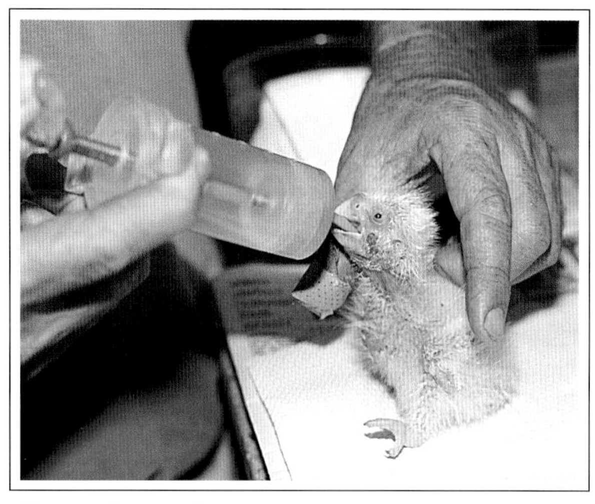

Above: Crop tubing is so easy. This White-tailed Black Cockatoo chick simply sits there while the crop is filled.

Below: You do not even have to take the chick out of the container.

The chief advantage of crop tubing is time saved. Everybody leads a busy life and crop tubing turns the feed process into an extremely quick procedure. In many cases it takes longer to prepare the mix than to actually feed the chicks. By using a 50ml Bovivet™ Plexi syringe sufficient formula can be drawn up to feed up to four chicks at a time, depending on the species and volumes being fed.

How long should it take to hold the chick, introduce the tube, fill the crop and withdraw? With a little experience about 10 seconds per chick, therefore, three to four chicks will be fed in less than a minute. With no cleaning up of the chick post-feed and exact knowledge of volumes fed, it is no wonder that more and more aviculturists are straightening out their spoons and putting them back in the cutlery drawer!

Such a method of feeding raises another concern. With the tube going down the neck and into the crop of several chicks directly after each other, isn't there a risk of cross-infection? Realistically the answer is yes, however let's keep it in perspective and look at ways of minimising the possibility. After all, in the majority of situations chicks are brooded together where they are constantly in contact with each others faeces, beaks etc. Providing the chicks are all healthy and gaining weight this is acceptable and

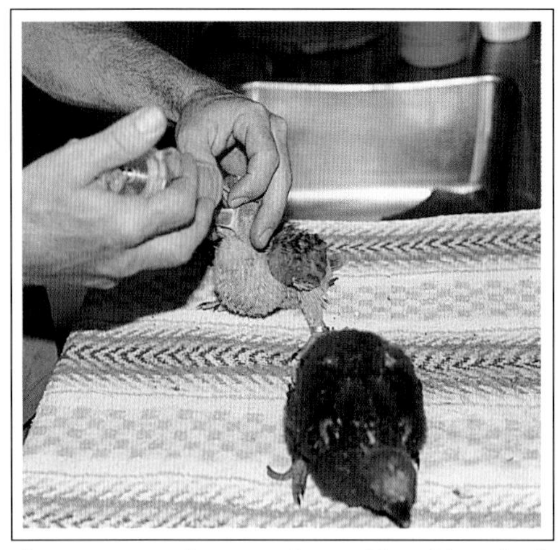

Left: Alexandrine Parrot being crop tubed.

Below: Crop tube feeding is quick. These six White-tailed Black Cockatoo chicks can be fed in less than four minutes.

the same applies to tubing. Should a chick develop an illness or begin exhibiting possible signs of infection then yes, it should be brooded individually and fed last with a separate tube. However, while the chicks are healthy it just is not practical to use a different tube for every chick. The majority of breeders feed any number of chicks with the one tube. Commonsense regarding treatment of the sick bird along with proper cleaning of the tube between feeds will leave little room for disease transmission. At the end of the day it remains a personal decision as to how far

Above: Yellow-tailed Black Cockatoo chick being crop tube fed.

Above right: Blue and Gold Macaw chicks showing all the signs of healthy growth.

Right: Blue and Gold Macaw chick being fed from a turkey baster.

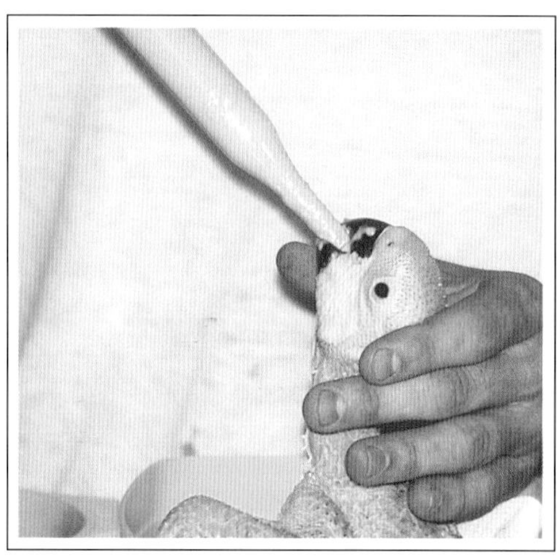

one should take the hygiene issue in this area and a compromise would be to have a separate tube per batch of chicks so that if there is an outbreak it is contained within a batch.

Total chick control is another distinct advantage of the tube and a sick/stubborn bird can be force fed. In many cases, the tube is the difference between life and death for the sick chick. The other benefit of the tube is that it generally produces birds that by nature are a little more aloof than spoon or syringe fed chicks. However, tameness of the bird is as much about the time spent with the chick as the method of feeding. Tubing being so fast, interaction time is minimal, so it is simply a case of spending time at each feed scratching and playing with the chick so that it begins to see you as a friend.

Tubing the weaning bird can be a problem as it develops its own personality and resists feeds, so this will be dealt with in the section on *Weaning*.

Above: Wipe clean the outside of the beak after each feed.
Below: Regularly clean old formula from inside the beak using a moist cotton wool bud.

FEEDING THE NEWLY HATCHED

So there is a newly hatched chick drying out in the brooder. It has been cleaned up and the navel area treated with Betadine™, what next? How soon should feeding begin? What should be fed? How much should be fed? How often?

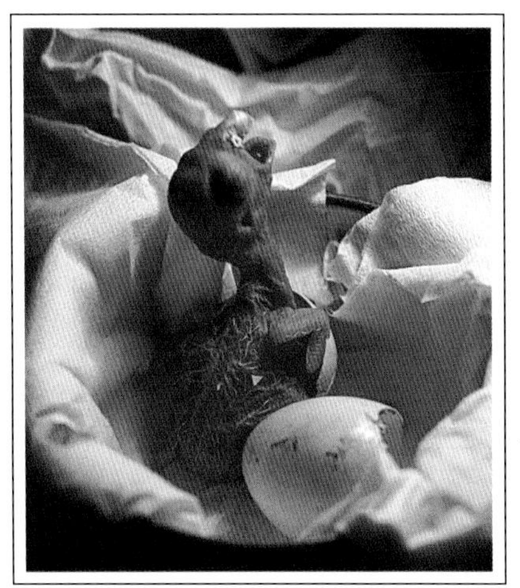

Most of the questions that confront the inexperienced have simple answers and providing the chick was a problem free hatch everything should proceed smoothly.

Small chicks have an incredibly strong desire to live and what they lack in size, they make up for in willpower. The important thing to remember here is that you are not dealing with a pin-feathered chick pulled from the nest and whose crop capacity, digestion rate and growth rate have

Left: Small chicks have an incredibly strong will to live. Here a newly hatched Major Mitchell's Cockatoo chick is up and looking for a feed.

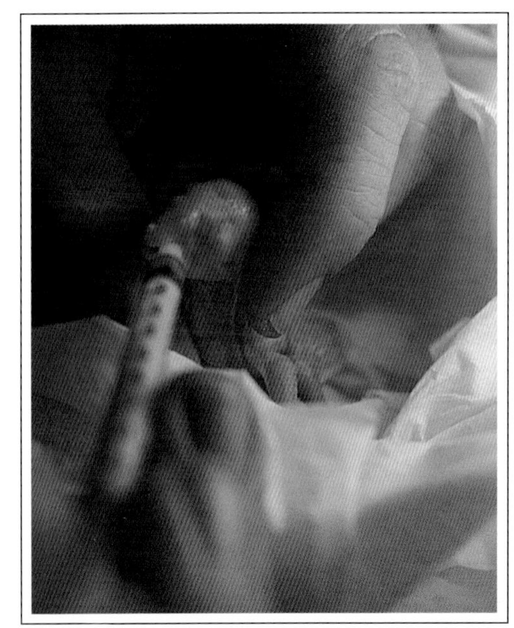

Left: Feeding by 1ml syringe allows volume control.

been established by the parents. With a newly hatched chick you are the parent and you must set that chick up for the growth phrase. The two important principles to practice at this time are:

- Rehydration
- Crop stretching

Carried out properly, you will produce a vigorous healthy chick with a fast moving crop that will power through the growth phase.

Rehydration

The majority of chicks hatch weak and in some cases dehydrated. Therefore they should be rehydrated by feeding an electrolyte solution with the addition of an energy source, before introducing handraising formula. It is important to understand that the body actually uses fluid to process food and unless the fluid deficit in the chick is replaced before feeding of formula commences, that chick will remain dehydrated as it begins digesting food. This results in a chick that never really reaches its full potential in terms of development and is more likely to experience problems. In severe cases of dehydration, where formula is introduced from hatch, the chick's body will simply draw the fluid out of the crop leaving the sediment behind.

There are several mixes that successfully rehydrate the newly hatched chick. One is the Ensure™ mix. Ensure™ is a nutritionally balanced powder for infants and is available from the chemist. Reconstitute the powder with boiled water (not boiling) according to the manufacturer instructions and then dilute 50/50 with lactated ringers (Hartmann's solution). Ensure™ provides the nutrition while the lactated ringer rehydrates the chick with electrolytes and replaces body salts. Other electrolyte mixes for rehydrating

Left: Rehydration mix for the newly hatched, Ensure™, spring water and Hartmann's solution.

Below: Feed under a heat source to maintain body warmth while outside the brooder.

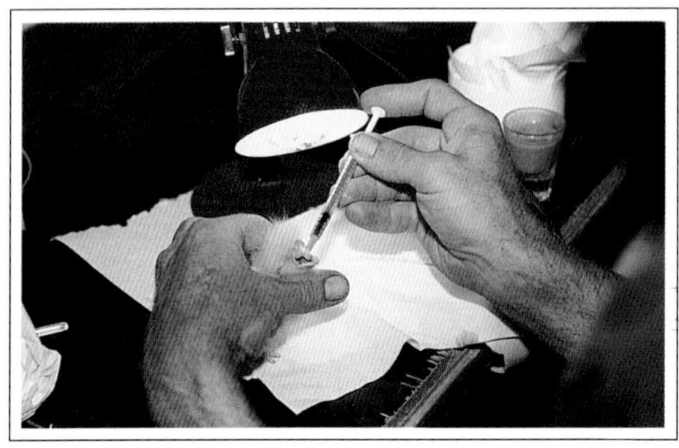

the chick are Gastrolyte™ available from chemists, and Spark™.

These fluid mixes should be fed for approximately the first 48 hours before introducing the handraising formula. Often a chick fed these fluids will exhibit a considerable weight gain in the first 24 hours, then very little more for the next one to two days. This initial gain, which can be as much as 6 grams in the Black Cockatoos, is a result of the chick's body tissue absorbing fluid and rehydrating.

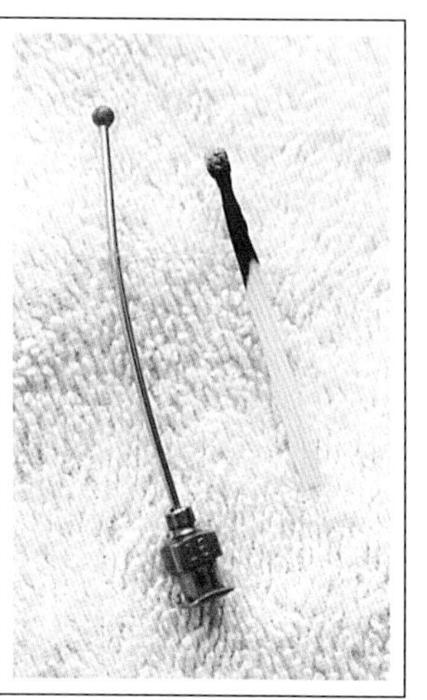

Above: A four day old Major Mitchell's Cockatoo chick with full crop.
Right: A very small crop needle to be used only in an emergency with a very small chick.

Most aviculturists leave the chick in the brooder six to eight hours after hatch before the first feed, allowing the chick to rest and dry out. If the hatching process was long and drawn out, or if it was a dry or assisted hatch and the chick is severely dehydrated (red, wrinkled, thin) then it should be rehydrated immediately. To place even very thin handraising formula into such a chick's crop will possibly overload the crop, causing it to shut down altogether. However, it is known that some aviculturists have fed thin formula to the chick from the first feed and produced quality chicks. Providing the chick hatched in excellent health and the formula was extremely thin eg: less than 10% solids, this can be tolerated by the chick, however, all chicks *will* do slightly better if rehydrated first, even if only for 12-15 hours. Even once the thin formula is introduced to the chick, it may be beneficial to continue adding rehydration fluids, eg. lactated ringers, to the formula along with normal mixing water, for up to a week.

Crop Stretching

Crop stretching is an important practice, particularly in the first few days of life and indeed throughout the entire growth phase. Put simply, it involves feeding slightly more in volume every feed, which will naturally result in more total volume fed every 24 hours. The principle is that a full, tight crop by virtue of inward and downward pressure, will speed up digestion. This is exactly what you want in the chick. High crop motility will see a healthy chick surge through the growth phase and peak around adult body weight, sometimes greater, for the species. This is the goal and crop stretching is how that is achieved.

Following is an example of the feeding regime of two White-tailed Black Cockatoos. Example A peaked at 732 grams and Example B peaked at 530 grams. The table shows where the reasons for this difference began.

WHITE-TAILED BLACK COCKATOO WEIGHT GAIN EXAMPLES

Peak Weights A = 732 grams
B = 530 grams

DAY	WEIGHT (grams)	GAIN (grams)	FEED TIME AND VOLUMES FED IN MILLILITRES																			TOTAL VOLUME FED IN 24 hrs
			6-7 am	7-8 am	8-9 am	9-10 am	10-11 am	11-12 am	12-1 pm	1-2 pm	2-3 pm	3-4 pm	4-5 pm	5-6 pm	6-7 pm	7-8 pm	8-9 pm	9-10 pm	10-11 pm	11-12 pm	12-1 am	
1																						
A	16	-		0.5		0.6		0.6		0.65		0.7		0.75		0.8		0.85		0.85		7.1
B	19	-		0.4		0.4		0.4		0.45		0.5		0.5		0.55		0.55		0.6		4.3
2																						
A	18	2		0.9		1.0		1.2		1.3		1.3		1.4		1.45		1.5		1.6		11.5
B	19	-		0.6		0.7		0.75		0.8		0.8		0.9		0.9		-		1.0		6.4
3																						
A	20	2		1.7		1.8		1.9		2.0		2.0		2.25			2.3			2.5		16.5
B	20	1		1.1		1.2		1.3		1.5		-		1.65			1.6			1.7		10.0
4																						
A	24	4		2.5		2.4			2.6			2.6		2.6			2.7			2.7		18.0
B	21	1		1.7		1.7			1.8			1.9		2.0			2.1			2.1		13.3
5																						
A	25	1		2.7			3.1		3.1			3.2			3.4			3.5		3.7		22.7
B	23	2		2.2			2.2		2.2			2.4			2.8			2.8		-		14.6
6																						
A	30	5		4.4			4.2			4.2			4.3			4.4			4.4			25.9
B	26	3		3.0			3.0			3.1			3.2			3.4			3.4			20.1
7																						
A	34	4		4.6			5.2			5.2				5.4		5.3			5.2			30.9
B	27	1		3.5			3.5			3.7				3.7		4.0			4.0			22.4

In Example A, note that almost every feed is increased slightly in volume. On Day 1, the first feed was 0.5ml and by the end of Day 1 the crop was holding 0.85ml. At the end of Day 2, the crop was holding 1.6ml, double the volume of Day 1, yet still emptying in the same two hour period. This is crop stretching at work - consistent volume increases have sped up digestion. Compare this to Example B, whose first feed on Day 1 was 0.4ml and last feed 0.6ml. At the end of Day 2 the volume fed was only 1ml. Compare also the differences in total fed per 24 hours. These differences in volumes fed continued for the entire process and the difference in peak weight between A and B demonstrates the benefits of crop stretching in producing a good sized chick in the end. These increases are in relatively large chicks. With smaller species, in the first few days, do not place too much importance in achieving a volume increase every feed during a day, or even every few feeds. One volume increase daily will suffice and providing the volume fed per 24 hours is on the rise then this is acceptable. Certainly, a normally developing chick should never be fed less than the previous day as this constitutes underfeeding. Good record keeping and volume control prevent this.

The first few days of life are a real window of opportunity, where you the parent, can set the stage for a quality chick to be produced, this being done by rehydration and crop stretching. Miss that calling and the result will be a chick that never quite made it.

The term crop stretching needs to be further defined. Notice that the volume increases are only slight, as it is very important not to overstretch the crop. Once formula or fluid has filled the crop and begins to sit in the lower neck area then no more should be fed. Overfeeding can be just as dangerous as underfeeding. To over feed, especially while on the rehydrating mix may see the chick aspirate as fluid flows back up the oesophagus and enters the windpipe. Crop stretching is not about putting such large volumes in the crop so that the muscle wall is pushed too far, losing elasticity. The problems associated with this are discussed in the *Troubleshooting* section.

The 1ml syringe through to the 5ml syringe are the suggested feeding instruments to be used for the first few days of life because they accommodate total volume control. If spoon feeding, volume control is very difficult, however crop stretching can still be performed by simply keeping the crop full at each and every feed. This way the crop will accommodate that little bit extra every feed and volumes fed will keep rising.

The First Feed

Most chicks have a feed response, albeit a weak one, from the first feed and are quite easily fed. Chicks that do not have a feed response initially, generally develop one after two to three feeds. Species vary in their feeding response. Major Mitchell's Cockatoos and cockatiels are excellent feeders from the first feed, whereas some Black Cockatoos may take up to three days to develop a feed response, and Eclectus at best have a weak feed response. Do not be too concerned over the chick that does not pump initially, it just means that feeding will be a little slower and tedious until the chick gains strength and develops a response, generally within three to four feeds.

To feed the chick, place thumb and forefinger either side of the beak and raise the head slightly. Dribble a small amount of formula onto the tongue and allow the chick to swallow before introducing the next drop. Do not place so much formula into the beak that it overflows as this increases the risk of aspiration when the chick tries to breathe before all the fluid is swallowed. Feeding the chick inside its brooding container is a little less stressful than being handled in and out of the brooder for each feed and also saves a little time. Some aviculturists prefer to crop tube feed with a very fine instrument until a feed response develops and if there are several such chicks this does save time. There is no harm in doing this, however it is a very delicate procedure and not recommended unless first experienced in the tubing of older chicks.

Feed young chicks under a heat source to keep them warm while outside the brooder. With the young chick being fed eight to ten times a day, if it is allowed to chill each time, then problems in terms of crop movement can develop. Wrapping the chick in a

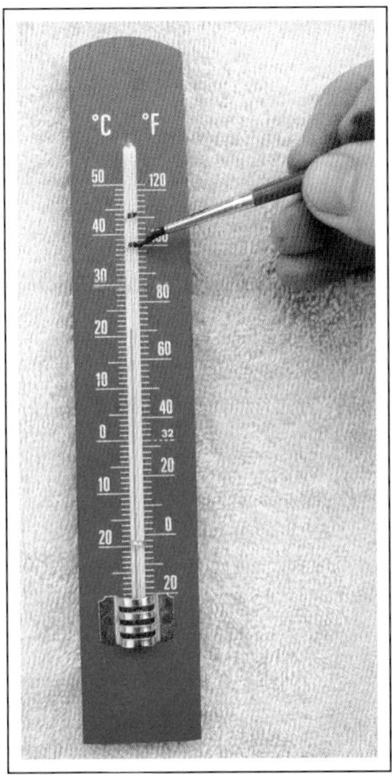

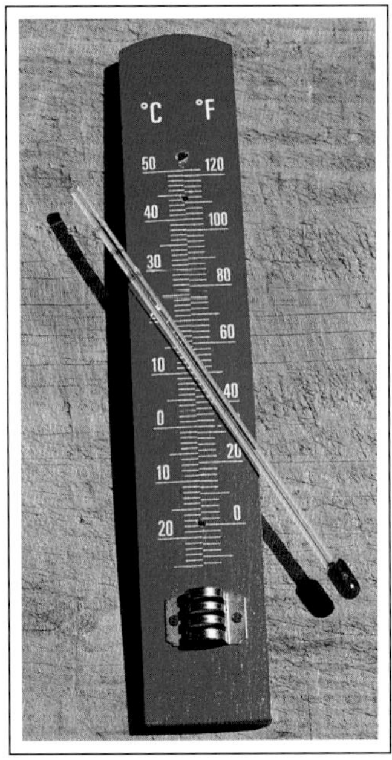

Left: Mark the temperature range with oil-based paint 37.7 - 43.3°C (100° - 110°F).
Centre: Remove the thermometer from its backing.
Right: Place thermometer straight into the formula for accurate measurement of temperature. Placing the formula inside a bowl of warm water helps maintain temperature.

tissue during feeding also keeps the chick warm and is recommended in the absence of a heat lamp. *Feed temperature for all chicks should be 40.5° - 41°C (105° - 105.8°F), however anywhere between 37.7° - 43.3°C (100° - 110°F) is tolerable.* The beauty of the syringe when feeding small chicks, apart from the advantage of volume control, is that the formula stays warmer much longer than on an open spoon. While feeding is slow in the early stages, the formula may be cooling too quickly on the spoon and the chick will refuse to feed. With the very small species it helps to file the tip of the syringe down to a blunt point so that it will fit into the beak.

Brood the chick upright if possible for the first week or so, particularly while on very thin mix. This will prevent the chick sleeping on a full crop and possibly forcing fluid back up the oesophagus causing aspiration. Do this by using a relatively small container and tucking tissues around the chick to keep its head elevated above the crop.

An interesting situation arises in the occasional chick, during the first few days of life, where during feeding it gulps down large amounts of air. When this happens the crop quickly balloons out and becomes distended while actually containing very little fluid or formula. Reasons for this remain unclear, however, it appears to be a physical dysfunction. There is no need for alarm, although this air needs to be removed if the chick is to be fed enough volume. Do this by lowering a fine crop tube or crop needle into the crop allowing the air to escape. Resume feeding and repeat the procedure if necessary until the crop is full of formula. Some

Right: The Creature Meter temperature probe.

aviculturists prefer to bypass the feed response by crop tubing at this stage and this unusual characteristic generally passes after a few days when normal feeding can resume.

Feed Intervals

Individual breeders feed their chicks at differing times however, the majority of aviculturists feed every 2-2 1/2 hours over an 18-20 hour period for the first few days before beginning to increase the intervals.

Assuming the chick hatches during the night, begin feeding fluids at approximately 7.00 am and go through until approximately 1.00 am, being over an 18 hour period. Whether to feed again between approximately 1.00 am and 7.00 am is an individual choice, depending on the chick's condition. For example, if the chick hatched around lunchtime or midafternoon it is suggested to feed every two hours through the first night to rehydrate and strengthen the chick as quickly as possible. Similarly, should the chick be very weak or dehydrated at hatch, it is wise to rehydrate every two hours for the first 24 hours.

How well the chick copes with a six to seven hour period of no feeding during the first few days depends directly on the management of that chick during the rehydration period. If the crop is moving well, volumes are increasing, the chick is healthy and strong and there is a nutrition source in the fluid (eg. Ensure™) then yes, the chick will tolerate seven hours between feeds without being compromised. The reverse also applies. Should the diet be inadequate or the chick is being underfed then it will need fluids somewhere in the middle of the early morning hours. Until experienced it is better to err on the side of caution and supply the extra feed at say, 3.00-4.00 am. Larger species, such as cockatoos, simply due to their larger body size cope better without this extra feed, than do smaller species.

There may actually be benefits in allowing the crop to sit empty for a period of time each night. An interesting occurrence has been observed in small chicks fed considerable and increasing volumes around the clock without a break. After two to three days of this regime the chick loses its feed response and will physically pull its head away when feeding is attempted. The crop also slows down and it appears almost as if the chick's body has been overworked by continual digestion and is trying to say, 'I need a break'.

As with handraising in general, what works for one chick may not suit another and every chick will be handled differently. A rule that does apply to the entire process until peak is - do not wait for the crop to totally empty before re-feeding during the feed period. The last portion of formula in the crop is extremely slow to move out and to wait for that to happen will actually result in underfeeding. Re-feeding on top of some of the last feed is recommended, however, not if more than 25% of the previous feed remains.

Introducing Formula

The change from rehydrating fluids to handrearing formula, and in fact any change in diet, is best done on an empty crop and after the extended night break. Introducing formula will see the crop slow a little as the body digests the food. This may result in the total volume fed per 24 hours remaining the same or even reducing a little for the first day of the changeover. Although this is fine, keep moving forward from that point. Feed interval increases are a natural progression once feeding formula, as the volumes are increased and the mix is thickened. By Day 7-8 the chick should be on three to four hour feeds. If by Day 7 the chick is still on less than three hour feeds, you are simply making work for yourself for no gain. However, try not to thicken the formula too quickly as the crop may slow down too dramatically. If in doubt err slightly towards a thinner mix.

Weight Gains

Handfed chicks almost always exhibit lesser weight gains than parent fed chicks in the first few days. There are several factors that contribute to this:

- Artificial diet
- Artificial brooding environment
- Lack of gut flora/enzymes from the parents
- Lack of parent-chick interaction/stimulation

For example, following is a table of a parent reared Yellow-tailed Black Cockatoo weight gains compared to three different handfed chicks reared by three different breeders.

Day	Parent Reared (grams)	Handreared (grams) (Three different breeders)		
1	24	22	24	25
2	32	27	27	31
3	40	27	30	38
4	60	27	32	37
5	78	29	36	39
6	88	33	42	44
7	106	38	49	55
8	111	44	55	65

With this in mind and depending on the graduations of the scales, a gain may not be recorded for up to three days. Should this occur do not be overly concerned, because the chick should stabilise and begin moving forward. However, be concerned if there is still no gain after three days of life or even worse, a loss. In fact, a loss anytime prior to peak should be of concern.

Temperature

The small chick, as with all forms of new life, is more susceptible to subtle environmental changes than older chicks. Three important aspects regarding temperature are:

1. Feed temperature range is 37.7° - 43.3°C (100° - 110°F) however, the closer to 41.1°C (106.0°F) the better. Some chicks are finicky and prefer it either slightly cooler or warmer, however to go outside these parameters is dangerous. The formula container can be kept warm during feeding by sitting it inside a pot of warm water. Do not guess the feed temperature with small chicks. Testing the temperature with the finger or on the wrist can be dangerous. Over time, these areas of skin will lose sensitivity and what feels to be warm formula may in actual fact, be too hot. Also, people have different perceptions to temperatures, therefore a thermometer is recommended.
2. Brooder temperature will directly influence progress. Incorrect temperature settings will result in slow crop and poor gains. Often a minimal adjustment in brooder temperature will see the chick improve.
3. Keeping the chick warm while feeding is important. The small chick will be in and out of the brooder many times in a day and if it chills every time it will become difficult to feed and develop problems.

GROWTH PHASE

Somewhere around Day 7 the growth phase begins for the vast majority of parrot species and as depicted in the *Goal* graph on page 37 it is about this time that weight gains should begin to dramatically improve. A chick that has been underfed thus far and not sufficiently rehydrated will have a somewhat slower crop and will struggle to reach its full potential during the growth phase. Also, these chicks are more prone to problems of one sort or another during growth and this is why it is important not to miss your calling with the chick in the first few days.

On the other hand, a fully rehydrated chick that was fed slightly increasing volumes each day will at this point be churning through the food and ready to develop into a robust, healthy bird with excellent gains. With feed intervals on the increase the growth phase is simply about feeding enough food to the chick so that it peaks somewhere close to and preferably over adult body weight.

Chicks are best brooded together at this stage and they really should do just three things, eat, sleep and grow!

Feed Instrument

For the first week of life it has been suggested that feeding be done by using a 1ml through to 5ml syringe, depending on the size of the chick, or by spoon, if that is preferred. Crop tubing has not been recommended because it is a delicate operation and bypasses the feed response, which during the first few days can be used as an excellent indicator of the chick's health.

Once into the growth phase however, all forms of feeding instruments are effective and several factors will influence the final choice, eg. the type of formula being fed, time available, number of chicks being fed and importantly, a method you feel comfortable with.

Crop Stretching

Crop stretching is important during the growth phase. Increasing volumes will keep the crop moving quickly and produce excellent growth. In the first few days the chick was fed slightly increased volumes, not only over most feeds, but every 24 hours the volume total was also higher. During the growth phase the volume increases will be a little more substantial though less frequent, allowing a chick to sit on a fixed volume for several days and still achieve maximum gains. Following is an example of feeding and weight records of a Major Mitchell's Cockatoo chick.

MAJOR MITCHELL'S COCKATOO CHICK

<table>
<tr><th>Day</th><th>Weight
(grams)</th><th>Gain
(grams)</th><th colspan="12">Feeds
(Time-Volume in millilitres)</th><th>Total Volume
Per Day (ml)</th></tr>
<tr><td>22</td><td>168g</td><td>16g</td><td colspan="3">7.00am
20ml</td><td colspan="3">12.00pm
20ml</td><td colspan="3">5.00pm
20ml</td><td colspan="3">10.00pm
20ml</td><td>80ml</td></tr>
<tr><td>23</td><td>172g</td><td>4g</td><td colspan="3">7.00am
20ml</td><td colspan="3">12.00pm
20ml</td><td colspan="3">5.00pm
20ml</td><td colspan="3">10.00pm
20ml</td><td>80ml</td></tr>
<tr><td>24</td><td>192g</td><td>20g</td><td colspan="4">7.00am
25ml</td><td colspan="4">2.00pm
25ml</td><td colspan="4">10.00pm
30ml</td><td>80ml</td></tr>
<tr><td>25</td><td>212g</td><td>20g</td><td colspan="4">7.00am
30ml</td><td colspan="4">2.00pm
30ml</td><td colspan="4">10.00pm
30ml</td><td>90ml</td></tr>
<tr><td>26</td><td>230g</td><td>18g</td><td colspan="4">7.00am
35ml</td><td colspan="4">2.00pm
35ml</td><td colspan="4">10.00pm
35ml</td><td>105ml</td></tr>
<tr><td>27</td><td>254g</td><td>24g</td><td colspan="4">7.00am
35ml</td><td colspan="4">2.00pm
35ml</td><td colspan="4">10.00pm
35ml</td><td>105ml</td></tr>
<tr><td>28</td><td>260g</td><td>6g</td><td colspan="4">7.00am
35ml</td><td colspan="4">2.00pm
35ml</td><td colspan="4">10.00pm
35ml</td><td>105ml</td></tr>
</table>

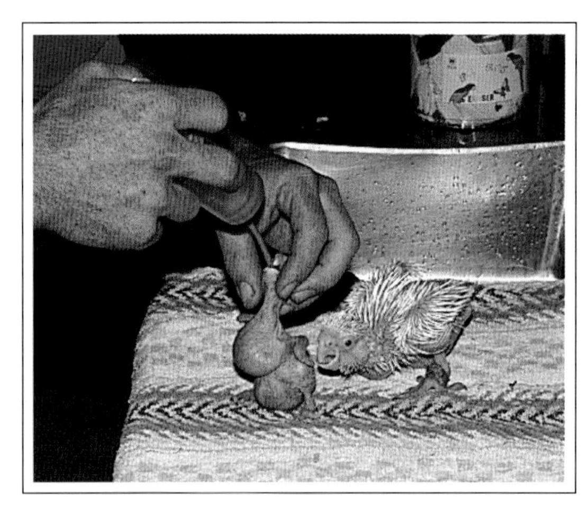

Left: Crop stretching continues during the growth phase.

It will be noticed that on Day 24 a feed was dropped and the chick went onto three feeds a day. Basically, all species will easily adapt to three feeds a day between Day 24-30, however factors that may prolong this drop to three feeds a day are: type of formula, the health of the bird and how effectively crop stretching was performed. To drop a feed and still maintain the same volume of formula entering the chick, it becomes necessary to increase the volume of the remaining feeds noticeably eg. 20ml up to 30ml. Increases in older chicks should not be more than 50% of the prior feed and less if the chick is younger.

It is important to progress to the next feed interval as quickly as possible, as to leave it too long will require too great an increase in the remaining feeds, potentially overstretching the crop. The other reason is that feeding a smaller amount more frequently will hold the chick back slightly in terms of weight gain. Major Mitchell's Cockatoos tend to be more susceptible to overstretching of the crop than other species and hence increases should be a little less each time.

Thickening the Formula

Somewhere between Day 7 and 16, depending on the species, the individual chick and the type of formula being used, the formula will be thickened to its maximum and the chick will remain on that consistency until weaned. While being careful not to rush the thickening process that began on Day 3 when formula was introduced, do not delay thickening unnecessarily either. Commonsense prevails, remembering that the thicker the food, the slower it passes. If the chick has a slow crop then the thickening process will be more gradual and if the chick is ploughing through the food then it can be sped up a little. Consult your records and if an increase in thickness saw a dramatic lengthening of emptying time then it was thickened too quickly. As already mentioned, adding peanut butter will slow crop motility, therefore it is not advisable to begin adding peanut butter and thicken the formula at the same feed.

Feed Intervals

Thus far, the chick began on approximately two hour intervals, nine to ten feeds a day. As volumes and thickness increase, work towards three feeds a day by Day 24-30.

Following are some general guidelines to work towards:

On Day 1	2 hour feeds	(9-10 feeds a day)
By Day 8	3½-4 hour feeds	(5-6 feeds a day)
By Day 14	5 hour feeds	(4 feeds a day)
By Day 24-30	8 hour feeds	(3 feeds a day)
At peak		(2 feeds a day)

Smaller chicks may be slower to reach four feeds a day however, any chick ten days or older need not be on less than four hour feeds (more than five feeds a day).

Accurate mixing of formula/water ratios becomes important once a routine pattern is established. If measurements are guessed then some feeds will be thicker, some thinner and this will affect crop emptying periods and gains. Using accurate measures of formula and water go a long way towards establishing predictable feed times and continuity.

As food sits in the crop it slowly begins to sour, slowing digestion. Fortunately food moves quickly enough through the healthy chick to prevent it souring too much, however each subsequent feed does tend to sour that little quicker because it has been added to the souring remains of a previous feed. To break the downward spiral most aviculturists practise allowing the crop to totally empty once in every 24 hour period, generally the period between last night feed and next morning, and this is strongly recommended.

However, the choice is individual. Some aviculturists, in order to achieve maximum weight gains, feed their chicks a greater volume at the last feed at night than during the day, which often results in formula still present in the crop next morning. Providing it is only small amounts and the chick is healthy, the fact that the crop never totally empties is of little concern. However, if the amount found in the crop each morning begins to increase then it is advisable to reduce the night feed volume to allow the crop to totally empty.

Weight Gains

The growing chick should gain some weight every 24 hours. The gains will be erratic and generally a large gain is followed by a small one. For example, a King Parrot may show a gain of 15 grams one morning and 2 grams the following morning. Records become invaluable during this growth period. Before handraising a new species, make the effort to obtain some weight gain charts from at least one other breeder, for comparison. If your chick's weight gains compare favourably with others, it suggests that everything is on track. Refer to weight gain tables, page 94. When comparing records establish whether they are of chicks handfed from Day 1 or pulled after a period of parent feeding. The reason for this as already discussed in the *Feeding the Newly Hatched* section, is that weights of a chick that received the parental advantage in the first week will be significantly greater than the handfed chick.

Yellow-tailed Black Cockatoo chicks at five weeks.

The laws of multiplication apply to the growing chick. If it fails to put on appropriate weight in the first week or so it will then fail to reach the average weight for that species at two weeks of age and so on. For this reason poor weight gains need to be addressed and corrected as quickly as possible before the chick falls too far behind. The bulk of the chick's weight gains occur in the first two-thirds of the growth phase. Daily gains tend to taper off as the chick heads towards peak and devotes much of its energy into feather development.

Another excellent indicator of the chick's general health is the physical examination. A chick with a well-rounded breast muscle is developing well, while a thin chick with a prominent keel bone is being underfed, on the wrong diet or suffering from an underlying problem. Be aware however, that some species do naturally develop thinner than others during handraising. Black Cockatoos for example, never seem to put on the bulk of many other species, yet develop into fine specimens.

Leg Bands

When chicks are brooded together it can become difficult to tell them apart, particularly if they are the same species and around the same age. A small packet of coloured plastic wrap-around leg bands will see up to ten chicks easily identified by colour and these bands can be adjusted to suit most leg sizes.

Long-term identification is also important as stud-books, genetics and selective colour breeding become an integral part of professional aviculture. The best type of ring is the stainless steel closed ring as opposed to a split or open ring. It is worth noting that aluminium closed rings are suitable for smaller parrots. The closed ring must be fitted relatively early in the growth phase while it will still slip over the toes and feet. By the time the chick peaks its feet have grown such that the ring cannot be physically removed. Depending on the supplier, rings will have various codes or numbers on them. A full history of the bird can be obtained by referring to the code in your record book. This is particularly useful to the buyer and lends credibility to the seller. Following are leg band sizes and approximate ringing age for various species:

Species	**Internal Diameter** (millimetres)	**Approximate Ringing Age** (days)
Green-winged Macaw	16mm	20-25 days
Scarlet Macaw, Blue & Gold Macaw	14.5mm	18-25 days
Black Cockatoos, Sulphur-crested Cockatoo	12-14mm	19-24 days
Eclectus, Major Mitchell's Cockatoo, Galah	10mm	12-14 days
Alexandrine, King Parrot	8mm	13-15 days
Large Rosellas, Superb Parrot, Indian Ringnecked Parrot	7mm	9-12 days
Sun, Jenday & Nanday Conures	6.5mm	14-16 days
Cockatiel, Western Rosella, Eastern Rosella, Kakariki, Plum-head, Slaty-head & Red-rumped Parrots	5.5mm	9-12 days
Lovebirds, Hooded Parrot	4.5mm	8-13 days
Neophema Grass Parrots	4mm	7-8 days

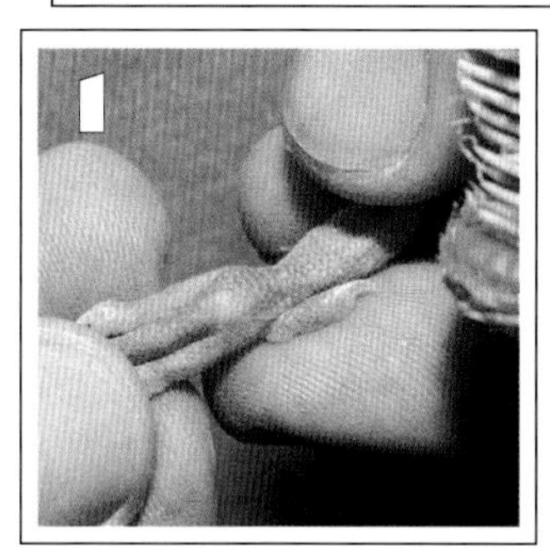

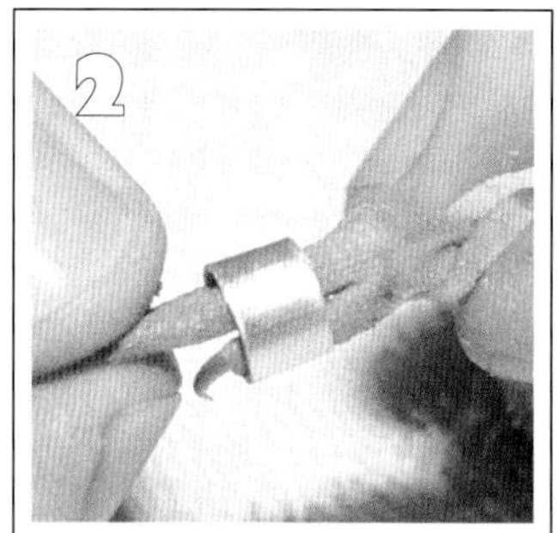

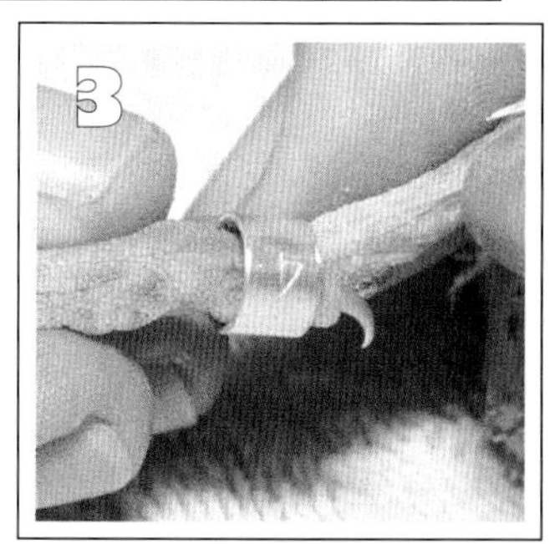

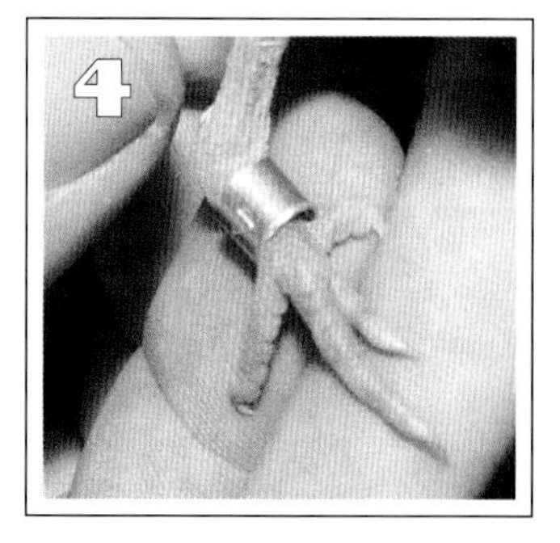

Series of photographs showing fitting of leg band.
1 and 2: The band or ring is placed over the three longest toes and moved down until it meets the fourth toe.
3 and 4: Gently pull the toe through the ring until it is located correctly on the chick's leg.

Above left: Closed rings are highly recommended for long-term identification.
Above right: Once chicks are fully feathered, they all look the same.
Right: Inexpensive, coloured plastic rings that wrap around the leg are excellent for identification during hand-raising.

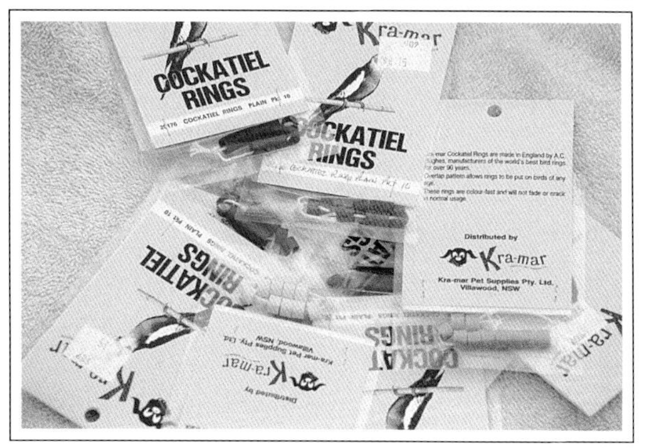

PEAK/WEANING PHASE

The next and final stage of the chick's development into an independent bird is the peak/weaning phase. The chick ceases to gain weight, stabilises for a period then begins to shed excess fat in preparation for its first flight. During this phase the chick begins to pick and nibble at food and slowly learns to break up and swallow its own food. The weaning phase passes uneventfully for most chicks, however, it can be a delicate phase and problems may arise, particularly with the larger species where weaning is a protracted event.

Recognising Peak

By recognising peak, in terms of management, you can prepare for changes in behaviour, feeding, housing etc of the chicks. If chicks are being weighed on a daily basis, then

Right: All these birds, except the small Black Cockatoo, are just about ready for the weaning cage and weaning food.

peak will be easily identified. Around adult body weight for that species, weight gains will taper off over a period of a week or so until one morning no gain, or even a loss, is recorded. This is *peak*. Following are the approximate ages at which some species peak, to be used as a guide if weighing is not taking place.

Species	**Approximate Peak Ages**
Green-winged Macaw	51-73 days
Scarlet Macaw, Blue & Gold Macaw	47-65 days
White-tailed Black Cockatoo	55-65 days
Red-tailed Black Cockatoo	55-65 days
Yellow-tailed Black Cockatoo	60-70 days
Eclectus Parrots	45-55 days
Major Mitchell's Cockatoo and Galah	40-45 days
King and Crimson-winged Parrots	35-40 days
Rosellas (except Northern Rosella)	20-25 days
Neophema Grass Parrots	22-30 days

These peak ages will be influenced to a degree by diet, volumes fed, whether there were health problems or the chick was a poor developer during growth.

Some individuals, especially the larger species, go through a second growth spurt anywhere up to two weeks after initial peak. Their weight, may in fact, climb higher than initial peak before dropping again. Generally, however, the chick peaks, remaining at that weight for a few days, then begins to lose weight.

Assuming the chick has been on three feeds a day, once it has definitely peaked it is safe and recommended to drop to two feeds a day. This will result in a drop in total volume fed each day, which is fine at this point in time. There are two behavioural changes that occur around this time and have caused many an anxious moment for the inexperienced (and some experienced too!).

- The chick becomes difficult to feed
- Regurgitation may occur

Feeding Difficulties

Most chicks will exhibit a noticeable decline in their feed response shortly after peak and in some cases this disinterest may even begin prior to peak. Expect this and remember that the chick, having stopped growing, will really only need maintenance volumes from now until it learns to feed itself.

The feed instrument being used may change hence the reason the serious handfeeder should be competent with spoon, syringe and tube. If using the spoon or syringe and the chick is refusing to eat (Eclectus love doing this) then, depending on how long this behaviour continues it may be wise to crop tube feed the bird once a day. This is to avoid excessive weight loss and dehydration until the chick experiences hunger and resumes feeding again, as most do. If tubing up until this point, those same chicks that happily swallowed the tube during growth may now become physically difficult to handle as their body tells them to reject food. The larger species may resist tubing so much at this time that they need to be wrapped in a towel for each feed. Unless competent with the tube it is suggested that you revert back to the spoon as most chicks will accept enough food during this phase and generally resume accepting the tube.

Many of the smaller species simply eat less per feed but remain relatively easy to feed and weaning is uneventful. It is the larger species that are more of a problem. It is only a phase though and virtually all chicks, after losing some weight and feeling hunger for a period, resume begging and pumping. It is important to introduce water to the chicks just prior to peak. Most learn to drink water long before they learn to eat and will cope much better with this low intake period if they can drink on their own.

Left: Around weaning, some of the larger species strongly resist crop tubing, requiring them to be wrapped in a towel for feeding. Alternatively, revert to spoon feeding.
Below: This phase soon passes and again they literally swallow the tube.

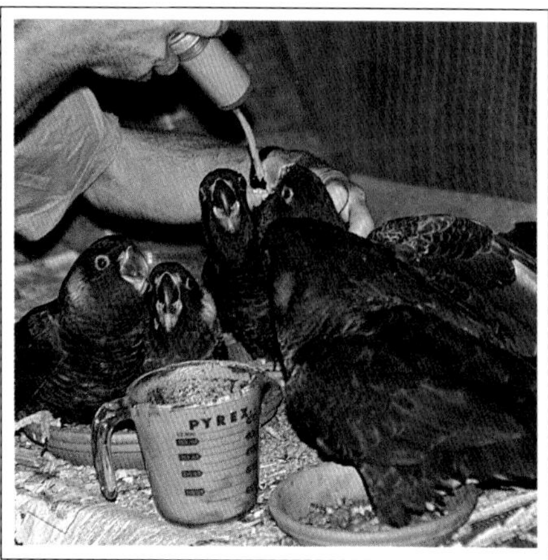

Regurgitation

Depending on the species and method of feeding, regurgitation will occur during weaning and to a degree this can be considered a normal part of the process. With some it will be simple head bobbing directly after each feed without actually bringing up formula, while in the worst case the bird will regurgitate the entire contents of the crop. It has been claimed that this trait is a result of the crop shrinking, however once this regurgitation phase passes, the crop is again able to accommodate large volumes. Rather, it appears to be a psychological response within the chick and is the body's way of saying, 'I'm trying to lose weight'.

Crop tube fed chicks are more prone to this behaviour than spoon fed chicks, simply because the spoon fed chick would not take more than it needed from the spoon in the first place. There is a degree of uncertainty as to just how much should be tube fed to chicks at this time and they often end up being fed more than they need, hence the regurgitation. Use records and volume control to find a volume that will be held down. Some chicks have a habit of holding the food down until the minute you leave the room, so regularly monitor for signs of regurgitation in your absence. Each feed, briefly examine the sides of the brooder/container for evidence of formula flicked around and if the chick is in a weaning cage, check the floor of the cage. This is relatively important if no longer weighing, because left unchecked long enough, there may be excess weight loss and dehydration.

If heavy regurgitation is occurring, it is time to reassess the feeding frequency and the volumes fed. Assuming the chick is at this point on two feeds a day, try reducing the volume fed by 25%. Alternatively revert back to three to four feeds a day at half the volume per feed. There are other things that can be done. The worst period is usually the few minutes directly after feeding. Divert the bird's attention until the initial desire to regurgitate passes by scratching it around the head and generally playing with it. Should the chick be in a wire weaning cage at this point, place it on the side wire directly after the feed. By using its beak (along with its claws) to hang on for dear life, it is not really in a position to throw up and by the time it finds its way to the perch, the urge has often passed. The fact that some chicks develop a habit of regurgitation the minute the feeder walks into the room, further suggests that the whole response is a psychological one.

Wherever there is constant, heavy regurgitation or difficulty in feeding, be it at weaning time or any other time, it pays to re-introduce electrolytes into the formula to help avoid dehydration of the chick. Also consider the possibility of an underlying medical problem and if in doubt, consult an avian veterinarian.

Weight Loss

The weight loss that begins shortly after peak is a natural occurrence and as long as it is within acceptable limits, there is no need for alarm. This weight loss is a necessary function in the wild as the chick prepares to emerge from the nest and takes its first flight. Its wing muscles at this stage are relatively undeveloped and to attempt its first flight while carrying unnecessary body weight would see it crash to the ground and fall prey to predators. Weight loss up to 20% of peak weight is acceptable, however any further loss needs to be addressed.

It is during this period of weight loss that the benefits of achieving maximum gains during the growth phase, become obvious. The healthy, well-rounded chick will easily tolerate up to 20% weight loss and remain healthy. However, the poor developer or undernourished chick can ill afford to lose any weight at all, let alone up to 20%. Again, this all gets back to how well the chick was set up during the first week of life.

Weighing, twice a week, still remains the best indicator of progress. The physical examination also remains important, a general feel of the breast muscle every few days will give a general indication of how much weight is being shed and how rapidly. Weighing of the weaning chick can be difficult because the chick is no longer prepared to sit placidly on the scales while being weighed, so it may be necessary to place it in a narrow cardboard box to restrict movement.

Weaning Cage

Once the bird is largely feathered and entering the weaning phase, it should be placed in a weaning cage. This is an important step in the weaning process. To take a fully feathered chick straight from the brooder container and place it in the aviary creates all sorts of problems.

The weaning cage is simply any type of wire cage or cabinet in which the bird can learn to perch, use its wings properly and begin to pick at food. Ideally, it should be big enough to allow free movement and exercise of the wings, even short flights, however small enough to keep it in relatively close contact with the food supply. An ideal weaning cage is one elevated on legs for ease of movement with a small door in the middle so the chick can be fed in the cage. During the early stages of introduction, place the perch/branch on the cage floor in a position where it will not roll around. There are two reasons for this. Even perching is a learning process and the chick will learn to step

Below: An excellent weaning cage. This perch is on adjustable brackets that allows the perch to be placed at floor level and gradually be raised as the chick develops. Also, the door allows for easy feeding of the chick while still in the cage and the cage itself is easily moved about.
Right: Two weaning Eclectus chicks.

onto the perch and grip much quicker this way than if the perch is suspended and it also keeps the food and chick close together. This way the chick is more likely to pick at the weaning foods more often, speeding up the weaning process.

Many of the smaller species will wean in a matter of weeks, while still in the weaning cage. However, the cockatoo family and other larger species of parrot will be moved to the aviary before completely weaned.

Weaning Foods

Use imagination when providing weaning foods. Some of the favoured foods are corn on the cob, peas, apple, carrot, lettuce, passionfruit halves, orange, wholegrain bread, Nutrigrain™ and silverbeet. Sprouted sunflower and lupins are also an excellent supplement for weaning chicks, however be sure to practice tight hygiene when sprouting. Provide a bowl of dry seed, as well as peanuts and almonds for the larger species such as Eclectus, macaws and cockatoos. It does not hurt to even introduce weaning foods prior to peak as some species begin to seriously pick and chew around late pin-feather stage. This early introduction may actually help usher in independence a little quicker. Lorikeet chicks learn to take their own nectar slurry and dry mix very quickly and may be totally weaned in a matter of days.

Many species can be encouraged to eat formula themselves shortly after peak by feeding them a spoonful then dipping their beaks into a bowl of formula. They soon connect taste with swallow and although their facial feathers become quite messy it does save time. The theory that crop tube fed chicks take longer to wean is really not correct as some tube feeders record extremely early wean times and at the end of the day it is more about how you wean than how you have fed the chick.

Above: Sprouted seed is an excellent weaning food, however practice tight hygiene using Aviclens™.
Below: Variety of fruit and vegetables suitable for weaning chicks.

Not long after peak, many species, especially the Black Cockatoos, go through a particularly strong curiosity phase and if this window of opportunity is used to its maximum advantage, the chick can be weaned in a minimum of time. The idea is to constantly maintain the chick's interest in the food bowls by presenting something new or different every few days, even changing the colour and size of the bowl. It is important to exploit this curiosity factor to its limit. The same food presented the same way every morning will soon bore the chick and its interest in the food will diminish, causing weaning to become the prolonged affair it often is.

Once the chick has been observed playing with the food or cracking seed, begin checking the crop before each feed. If food is present in the crop cut back the formula accordingly. In fact, if the crop is at least half full, a feed may not be necessary at all. Weaning is a tricky time because to overfeed may well keep the chick from beginning to find its own food while underfeeding may stress the chick forcing it back to complete dependence. Many people think that the hungrier the chick is, the quicker it will learn to eat itself, however starving can often have the opposite effect. After all, if the chick has not yet learnt what all that food in the bowls is about, how will it support itself? At the end of the day weaning is a trade-off period, where the more it eats itself the less it receives from the feeder.

When dropping from two to one feed a day, drop the morning feed. The chick will be slightly hungry during the day encouraging it to begin finding its own food. It pays to monitor the chicks weight by weighing every three to four days for up to three weeks after the very last feed, to ensure that it is in fact eating enough food to at least maintain weight.

Selling Chicks Unweaned

The selling of chicks during the weaning phase is risky and inadvisable. Many a weaning chick has been sold by an aviculturist with little handraising experience to a buyer with even less and the result is a bird that suffers and some even die. Weaning is a delicate phase and changes in feeder or environment can be all it takes to stress the chick to the point where it refuses to eat. What many novice breeders fail to fully understand is that weaning is a gradual decline in handfeeding as the chick increases its own intake. To cease handfeeding before the chick is totally self-sufficient will often cause a chick to revert back to complete dependence on the feeder. There are horror stories around of galahs and other pet cockatoos wasted away and begging for food at six to eight months of age, simply because the owner bought the bird unweaned and was oblivious as to the correct way to feed and wean such a bird. Fortunately, there are laws in place in several states that prohibit trade in dependant chicks for this very reason. At the end of the day the onus is on the individual selling the chick, to ensure if it is not weaned that the new owner fully understands what is involved and to follow up on progress. Some species, particularly cockatoos, have been sold totally weaned then gone back to begging and dependence after the trauma of transport to a dealer's premises or another breeder's aviaries. It is worth noting that some species, particularly the larger ones, remain susceptible to regression for some time after being weaned. Following is a table of approximate ages at which chicks should be weaned.

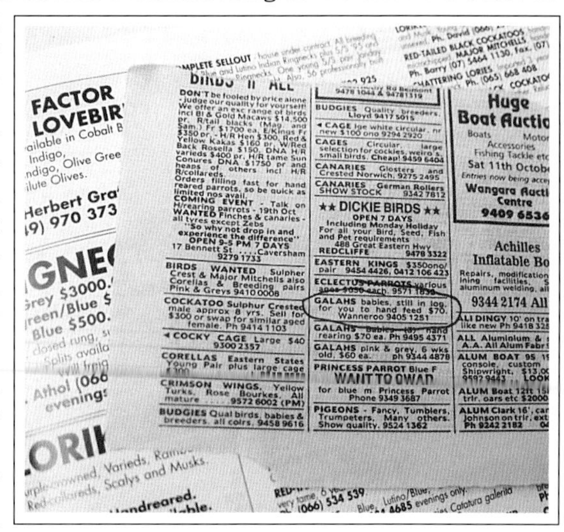
FACTOR LOVEBIR
Herbert Gra
(9) 970 373

DON'T be fooled by price alone - judge our quality for yourself! We offer an exc range of birds incl Bl & Gold Macaws $14,500 pr. R/tail blacks (Mag. and Sam.) Fr $1700 ea, E/Kings Fr $350 pr, - H/R Hen $300, Red & Yellow Kakas $160 pr, W/Red Back Rosella $150, DNA H/R varieds $400 pr, H/R tame Sun Conures DNA $1750 pr and heaps of others incl H/R R/collareds.
Orders filling fast for hand reared parrots, so be quick as limited nos avail.
COMING EVENT - Talk on H/rearing parrots - 19th Oct
WANTED Finches & canaries - all tyres except Zebs
"So why not drop in and experience the difference"
OPEN 9-5 PM 7 DAYS
17 Bennett St .... Caversham 9279 1733

BIRDS WANTED Sulpher Crest & Major Mitchells also Corellas & Breeding pairs Pink & Greys 9410 0008

COCKATOO Sulphur Crested male approx 8 yrs. Sell for $300 or swap for similar aged female. Ph 9414 1103

COCKY CAGE Large $40 9300 2357

CORELLAS Eastern States Young Pair plus large cage

CRIMSON WINGS, Yellow Turks, Rose Bourkes, All mature .... 9572 6002 (PM)

BUDGIES Qual birds, babies & breeders, all colrs, 9458 9616

BUDGIES Quality breeders, Lloyd 9417 5015

CAGE lge white circular, nr new $100 ono 9294 2920

CAGES Circular, large selection for cockies, weiro's, small birds. Cheap! 9459 6404

CANARIES Glosters and Crested Norwich, 9275 2495

CANARIES German Rollers SHOW STOCK 9342 7812

★★ DICKIE BIRDS ★★
OPEN 7 DAYS
Including Monday Holiday
For all your Bird, Seed, Fish and Pet requirements
488 Great Eastern Hwy
REDCLIFFE 9478 3322

EASTERN KINGS $350ono/pair 9454 4426, 0412 106 423

ECLECTUS PARROTS various

GALAHS babies, still in log, for you to hand feed $70. Wanneroo 9405 1251

GALAHS Babies (8) hand rearing $70 ea. Ph 9495 4371

GALAHS pink & grey, 6 wks old, $60 ea. ph 9344 4878

PRINCESS PARROT Blue F
WANT TO SWAP
for blue m Princess Parrot Phone 9349 3687

PIGEONS - Fancy, Tumblers, Trumpeters, Many others. Show quality. 9524 1362

RED-TAILED BLACK COCKATOOS
MAJOR MITCHELLS
Ph. Barry (07) 5464 1130, fax (07)

Huge Boat Auctio
Boats Motor Accessories Fishing Tackle etc
Sat 11th Octob
Wangara Aucti Centre
9409 6536

Achilles Inflatable Bo

Right: Selling of chicks before fully weaned has its problems and both buyer and seller must know how to handle the weaning chick.

Average Weaning Age of Various Species

White-tailed, Red-tailed and Yellow-tailed Black Cockatoos	120-150 days
Green-winged Macaw	110-120 days
Long-billed Corella	100-110 days
Scarlet Macaw, Blue & Gold Macaw	85-95 days
Galah, Major Mitchell's Cockatoo, Short-billed Corella	80-90 days
Eclectus Parrots, African Grey Parrot	80-90 days

King and Crimson-winged Parrots, Indian Ringnecked Parrot and Sun Conure	70-85 days
Princess Parrot, Red-rumped Parrot	40-50 days
Cloncurry Parrot	45-50 days
Rosellas	40-50 days
Neophema Grass Parrots	35-45 days

WEANING BLACK COCKATOOS

An increasing number of Black Cockatoos are being successfully bred each year in Australia as their popularity increases and management skills improve. Black Cockatoos have a few behavioural traits common to all members of the family. Some hens lay their eggs on the ground, many pairs hatch two eggs and invariably let the second chick die and some hens neglect their chicks during raising. These habits, combined with the fact that they are reliable relayers, if the first clutch is pulled early, has resulted in a large proportion of chicks handraised. This is great for aviculture, however the handraising of these birds is a prolonged affair and not without its share of attendant problems. For this reason it is felt appropriate to devote some text specifically to the weaning of Black Cockatoos in the hope of contributing to easier and earlier weaning.

Above: Peas and corn are excellent Black Cockatoo weaning foods.
Below: It is surprising how many Black Cockatoos wean themselves on plain canary seed, despite it being a small seed.

The chick will easily tolerate three feeds a day around Day 26-30, providing the chick is developing normally and has good crop motility. Once the chick has peaked it is then recommended that a feed be dropped shortly thereafter. The morning feed is then reduced gradually so that somewhere around Day 80-90 the weaning Black Cockatoo is on one feed only, at night. It is around this age that the chick goes through that particularly strong phase of picking and swallowing weaning food which lasts up to several weeks, before subsiding somewhat. If the chick was still on two feeds a day at this point, it may pass through this particularly strong curiosity phase without experiencing a little hunger and feeling the need to find its own food. As a result, the opportunity of early weaning may be missed. Place in a weaning cage and follow the weaning procedures outlined previously. Monitor weight loss and if losing weight too fast then ease up on the food reduction and add a little

Above: Peanuts and almonds for the weaning Black Cockatoo. Sunflower seed is necessary, however limit the amount eaten as an addiction may develop.
Below: An adult Yellow-tailed Black Cockatoo that cannot wait to be served his mealworms. This passion is the result of early introduction.

more peanut butter. Some Black Cockatoos will lose up to 25% of body weight and go surprisingly thin, yet wean successfully. Such extreme weight loss is not desirable and should be avoided if at all possible.

The type of weaning foods presented has a major influence on how quickly the Black Cockatoo is weaned. Offer a variety of foods as discussed under *Weaning Foods*. The addition of a separate bowl of plain canary seed is suggested. Surprisingly, this seed is often the first seed Black Cockatoo's learn to eat. Weaning a Black Cockatoo purely on sunflower seed is to be avoided because once it develops a taste, it is addicted and is an extremely difficult habit to break. However, do not totally exclude sunflower seeds from the weaning chick's diet as this family need this oily seed to maintain bulk during the weaning phase. Include with a variety of other seeds and soft foods.

Livefood is an important aspect of the Black Cockatoos diet and there is no better time than weaning to introduce mealworms. Break open the worms, expose the juices and handfeed them to the chick until they learn to pierce them with their beaks. Whole mealworms take on the texture of a piece of rubber and while the inquisitive chick gets hours of enjoyment from mealworms wriggling around in its mouth, it may not actually recognise them as a food source.

Perching and learning to fly and land are major learning events to the young Black Cockatoo and for that reason it is recommended that after leaving the brooding container, they spend some time in a smallish weaning cage learning to perch and flap properly before moving into an aviary where they can fly. Only when they can safely land and perch is it recommended to place them in a larger flight.

Black Cockatoos have been known to wean totally, remain completely independent for up to a couple of months, then, for various reasons, regress to total dependence again. Any chick that goes light or resumes incessant begging anywhere up to 12 months of age may be having trouble supporting itself and slowly going downhill. Generally speaking, Black Cockatoos should be weaned totally between 120-150 days (four to five months). There are reports of White-tailed Black Cockatoos weaned at 90 days, however it is suggested that this is a little young and may be pushing the chick too hard. On the other hand, nine to ten months is far too long for the healthy chick apart from the occasional 'sook' that has had problems along the way. Another result of this

Cockatoos are particularly prone to imprinting, where, instead of being affectionate to their mate (left), they are more attracted to the keeper (above). The outcome can be poor breeding results especially if the cock is the imprinted bird.

Left: House the weaning Black Cockatoos in a smallish outside aviary where they can acclimatise and have easy access to weaning foods.

Above: Time to move into the weaning cage.
Left: Due to the extended weaning time of the Black Cockatoos (120-150 days), handfeeding will continue once in the final aviary.

extended period of dependence on the feeder is imprinting, particularly where the spoon is being used and this can have an adverse effect on the breeding ability of some birds. Imprinted cocks become quite difficult to pair up because they are more interested in their keeper than their partners. To avoid this, spend as little time as possible with the chick and use the crop tube feeding method.

PULLING CHICKS

Above: *One day old Eclectus Parrot chick. When inspecting the young chick look for a healthy, strong chick with good pink skin colour.*

There are many reasons for pulling one or more chicks from the nest. Poor parenting, loss of one of the parents, a large clutch or to increase production are just a few. When pulling, it is recommended that the chick, as with the egg, be left under the hen for at least the first week, if at all possible. This way the chick will have received a good dose of natural gut flora from the parents and all the hard work involved in setting up crop motility and growth rate will be done for you. To pull the chick once in the growth phase, means that from then on it is simply a case of feeding the chick increasing food volumes until peak.

Never underestimate the parental advantage.

Inspecting the Young Chick

Many an aviculturist in peering into the nest to examine the condition of a young chick and being unsure of its true status, has erred on the side of caution and pulled it. The question is, 'How do I know the young chick is doing okay?'. Although every situation is different, there are several areas to consider when assessing the chick, which when addressed may save unnecessary pulling. Those main areas are:

- Weight gain
- Parental behaviour
- Visual examination
- State of the crop

Above: Northern Rosella hen leaving her nest. A good sign of chick progress is the activity of the parents.

Below: Be aware that some species eg. Hooded, Golden-shouldered and Plum-headed Parrots, have a habit of ceasing night brooding early. Seen in this photograph is a Hooded Parrot hen.

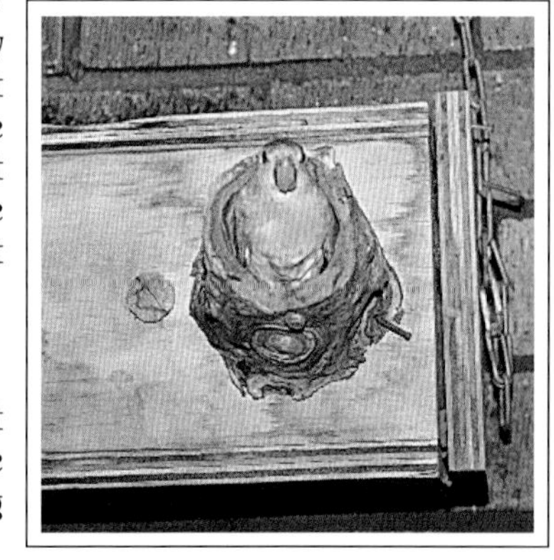

Weighing

As in the nursery, the one best guide to a chick's progress in the nest is weight gain. With a portable set of digital scales and an efficient inspection hatch, it is only a matter of minutes to weigh the chick each morning. This is more practical with the larger species where distinct gains can be recorded on a daily basis and where there is easy access to the chick via an inspection hatch. It is particularly successful with the cockatoo family, however is not suggested with Eclectus Parrots. Eclectus hens are particularly intolerant of interference with the nest and chicks. Hens have been recorded diving into the nest and attacking a chick directly after a nest inspection.

Parental Behaviour

The behaviour of the parents is often an excellent guide to progress without actually examining the nest. Generally speaking, providing the hen is sitting

Above left: An inspection hatch at nest level is essential. This Eclectus Parrot raised a Major Mitchell's Cockatoo from the egg and could be inspected daily.
Above right: Avoid pulling a chick with a crop full of hard food if possible. This is a parent-fed ten day old Hooded Parrot.

tight, the cock is feeding the hen on and outside the nest and large amounts of food are being eaten, this augers well for the chick's progress. In the first week of life the hen should only be off the nest for brief periods of time. As the chick ages, the hen will brood less during the day. Black Cockatoos are the exception to this and often a hen will cease daytime brooding as early as Day 3-5. Depending on the prevailing climate, do not be overly concerned with this as Black Cockatoo chicks are heavily downed and can tolerate quite cool conditions. Brooding at night still remains essential at this early stage of the chick's life and where the hen is an unproven mother it pays to observe her movements around dusk to ensure she enters the nest for the night. This is particularly important with species that are renowned for early cessation of night brooding, such as Hooded and Plum-headed Parrots.

Visual Examination

When inspecting the young chick in the nest, commonsense prevails. Look for a healthy chick with good skin colour. Any chick that is excessively red or pale or lying on its side may be in need of help, however be aware that lorikeet and conure chicks have a habit of sleeping in all manner of positions, including on their sides and are actually doing fine.

State of the Crop

Just because a young chick has an empty crop does not mean it is in trouble. Some hens feed small amounts often during the first two to three days and an empty crop alone does not warrant pulling. Providing the parents are behaving appropriately and the chick looks healthy and strong, do not be too concerned about the empty crop. Simply check it again later in the day. Many a chick found with an empty crop more than once in the first few days of life has grown into an excellent specimen.

On the other hand, nor is a young chick with a full crop necessarily doing fine. Often when a chick is found dead in the nest the breeder assumes, based on the presence of food in the crop, that the chick must have been doing fine right up until it mysteriously and suddenly died. What is possible here is that the chick was in trouble for days prior to death and that food in the crop may be old food that stopped moving out of the crop long before actual death. This is why the status of the crop should only be one aspect of a much more extensive physical examination.

Should a first time mother let her chicks die, do not immediately assume that she will

Left: Large clutches are more likely to contain a neglected chick. Examine carefully.
Below: Yellow-tailed Black Cockatoo chick at six days. It can be difficult to tell the true condition of a very young chick, especially if heavily downed.

always be a poor parent. There is a degree of learning for the first timer and she may well lose a chick or two before improving her nurturing skills. By automatically pulling all future chicks as soon as they hatch, based on a first time failure, the hen will never be given the opportunity to gain experience. Eclectus are again unusual - it appears to be almost expected to lose a chick first time yet thereafter raise every chick. Given a second chance the vast majority of hens prove themselves.

Problems in the Nest

A few problems can arise in the nest to warrant the pulling of one or all the chicks. Be suspicious of chicks continually calling from the nest, as this is a classic sign of underfeeding. Upon examination, should the chicks be on the thin side, then this is the most likely cause of their distress, however keep in mind other causes such as worm infestation or an underlying infection. There are many facets to successful breeding and often poor quality chicks are the result of inferior management on the aviculturist's part. Poor hygiene in the aviary, poor worming practices, poor diet, improper nesting material, poor nest protection from the elements, no vermin control etc. are all keeper based problems and it often pays to look beyond the hen for the answers.

Worm infestation must be considered as a possible cause of underweight chicks that continuously call from the nest. If intestinal worms are suspected, treat such chicks via the formula once in the nursery. Your avian veterinarian should be consulted for recommended dose rates and procedures.

Where large clutches are involved, it is not unusual to find one or two chicks, generally the youngest, falling behind the rest in development. Often these chicks huddle in the middle of the clutch and may be missed in a casual nest inspection. There are two schools of thought in such a situation, some breeders choose to pull a couple of the better developed chicks to allow the neglected ones to receive more attention, while others pull the weaker chicks, feeling they have a better chance of recovery in the nursery. If it is decided to pull the problem chick (which is suggested as the better way) be aware that the reason for its underdevelopment may in fact be that it is harbouring a health problem that may be transmittable to other chicks in the nursery.

Should you arrive home from work one afternoon to find the hen out on the perch and a cold lifeless chick in the nest, do not give up hope. Many an assumed 'dead' chick has responded to warmth. Place the chick in a dish lined with tissues inside a brooder at approximately 35°C and wait for it to show signs of life as it warms up. Once warm and moving it will need emergency rehydrating fluids eg. Hartmann's solution, Gastrolyte™, Spark™ or glucose and water, and in such a situation administering by tube becomes a lifesaver. Using a small tube, feed small amounts until the chick regains strength, then introduce very thin formula with extra apple sauce.

Best Time to Pull

The best time to pull a chick, should you have the choice, is around the middle of the pin-feathering stage. Leave it too late and there may well be difficulty in teaching it to accept feeding and handling. Pulling chicks around fledging time is not a good practice because by that stage many chicks have a strong fear factor. Combine this with the natural desire to reject food at this time, as well as the stress of a new environment and there may be excessive weight loss and dehydration.

The best time to pull chicks is at pin-feather stage, as seen in these two Major Mitchell's Cockatoos.

Pull chicks early in the morning when the crop is empty. Sometimes the parents are active surprisingly early and depending on what time you rise, the chicks may have already been fed. A way around this is to remove the food supply from the aviary last thing the previous night. The problem with pulling a chick that has a crop packed with seed and pieces of fruit is that due to the usual stresses, the crop will move slower for a couple of days. A full crop of solids may in fact stop altogether during this period and you head down the path of sour crop and manual emptying. If the solids are slow to move, tube in fluid with some glucose and apple sauce and give the crop a massage two to three times a day to break up the contents.

Regardless of the age of the chick when pulled, begin it on a thinner formula than normal for its age. The crop is slower at this point and full strength formula should never be placed in a slow crop, whatever the cause may be. Add extra apple sauce and as the crop motility resumes during the next 24-48 hours, thicken the formula. Keep up the volumes and remember that a full tight crop will speed up digestion.

With spoon or syringe feeding it may take a day or so until the chick regains its natural feed response. Be patient and careful with feeding in the meantime, allowing the chick to swallow the formula as it is administered. If tubing, the chick may struggle a little for a day or so, however by offering a spoon full first this will loosen the chick up.

Should you be tubing for the first time and have no records available the question arises, 'How much do I feed the chick?'. Good question. Following is a record table of an Eclectus chick pulled at Day 22. Looking at the table, the idea is to initially feed an amount that is safely on the side of less than normal for the age of the chick and then increase to maximum crop capacity quite quickly.

Right: Glucose, an excellent emergency supplement if Spark™ or lactated ringers are not available. Dilute in water and feed at 41°C.

ECLECTUS *E.r. polychloros* CHICK RECORD TABLE

Day	**Weight** (grams)	**Gain** (grams)	**Feeds** (Time-Volume in millilitres)				**Total Volume Per Day** (ml)
22	170g	12g	E 7.00am 12ml	E 12.00pm 15ml	NQE 5.00pm 17ml	NQE 10.00pm 18ml	62ml
23	176g	6g	E 7.00am 20ml	NQE 12.00pm 20ml	E 5.00pm 22ml	NQE 10.00pm 22ml	84ml
24	186g	10g	E 7.00am 24ml	NQE 12.00pm 24ml	NQE 5.00pm 24ml	NQE 10.00pm 24ml	96ml
25	200g	14g	E 7.00am 24ml	NQE 12.00pm 24ml	NQE 5.00pm 24ml	NQE 10.00pm 24ml	96ml
26	216g	16g	E 7.00am 24ml	NQE 12.00pm 24ml	NQE 5.00pm 24ml	NQE 10.00pm 24ml	96ml
27	235g	19g	E 7.00am 25ml	NQE 12.00pm 25ml	NQE 5.00pm 25ml	NQE 10.00pm 25ml	100ml
28	253g	18g	E 7.00am 33ml	NQE 2.00pm 33ml	NQE 10.00pm 34ml		100ml
29	264g	11g	E 7.00am 36ml	NQE 2.00pm 36ml	NQE 10.00pm 36ml		108ml
30	285g	21g	E 7.00am 36ml	NQE 2.00pm 36ml	NQE 10.00pm 36ml		108ml

E = Empty NQE = Not Quite Empty

Knowing that a chick of this age would easily cope with four feeds a day, the goal was simply to find the volume that would see the crop emptying in approximately five hours. On the first day (Day 22), erring on the side of caution, a thin mix of 12 mls was fed at 7.00am. The crop was empty at 12.00 pm, so clearly the volume could rise. At 12.00 pm, it was fed 15mls which left the crop not quite empty at 5.00 pm. This is good because an acceptable crop volume is established after just two feeds. At 5.00 pm the chick is fed a little more, 17mls and again produces a not quite empty crop at 10.00 pm. Remember that you should be feeding on top of a little food from the previous feed.

Day 23, the formula is thickened and the volume again increased. The chick has had a day to stabilise and adapt to a new diet and feeder. It is now time to settle on a set volume and keep up the crop stretching procedure until peak, as shown in the next seven days of the table. Depending on the crop movement and how well the chick adjusts to the changes, full strength formula can be fed on the second or third day.

Day 28, a feed is dropped and the chick is now on three feeds a day. The volume per feed now jumps from 25ml to 33ml so that there is no reduction in the total millilitres fed per 24 hours. The *Total* column should always be on the rise to achieve maximum growth potential and avoid underfeeding. It is important to remember that if the chick has been pulled due to parental neglect or underfeeding then its crop capacity will be less than normal for its age. Always lean towards a lower volume first feed to visually assess its true capacity. Following is a guide to maximum crop capacity for some species during handraising. These maximum volumes, if reached, will not be until middle-late growth phase.

Neophema Grass Parrots	10ml
Cockatiel	15ml
Indian Ringnecked Parrot	10-20ml
Rosellas	20-22ml
Eclectus	50ml
Black Cockatoos	80ml

Possible Long-term Consequences

There is no doubt that the incidence of eggs/chicks being pulled from the nest through choice is rapidly rising as the demand for pets increase and more breeders attempt to extract a financial return from their hobby. With such a trend arises the question, 'What impact will this practice have on the long-term breeding viability of captive stock?'.

This is an important issue in light of the fact that the day is already here where aviculture holds only limited numbers of some species, yet more than remain in the wild. The Spix's Macaw, Echo Parakeet and Golden-shouldered Parrot are such examples and there are many more. With these birds it becomes crucial, as aviculture enters the next century, that future captive generations retain the reproductive skills and desire of their wild counterparts.

Many aviculturists now push their breeders for several clutches of eggs each season and depending on how early in the season laying began and the quality of the season itself, many species are almost guaranteed to produce at least two and up to four clutches of eggs per year. This is great in terms of increased production but puts an incredible strain on the resources of the hen, particularly if the diet is a poor one. Continue this practice for several years and the result may be a hen that no longer has the stamina to continue as a layer, nor sees any point in doing so because the eggs continually disappear. Two practices are suggested here, restrict the hen to two clutches per season and three if the species is an all-year round breeder and secondly, allow the birds to raise at least one chick each year. Diet becomes critical where the hen is a relayer and must be balanced, varied and fresh. Calcium supplementation is recommended for all layers prior to the breeding season and becomes essential if more than one clutch is anticipated.

The possibility needs to be considered that if eggs are pulled for long enough, that, when or if the hen has the chance to raise her own chicks, she will have lost her rearing skills. Should this occur then effectively what is created is a pair of birds that are of little value to any breeder not prepared to handraise. True, nurturing skills in the parrot are largely instinctive, however there is undoubtedly a degree of learning and maintenance involved and the parents should be allowed the privilege of hatching and raising their own chicks from time to time.

This pulling of successive clutches of chicks within a season and every year not only has a direct effect on some breeders but may also have an as yet unrealised impact on the young. Concerning the parents, loss of a clutch of chicks can be highly traumatic, particularly to the hen and extended depression and moping has been observed in larger species such as macaws and Black Cockatoos after losing chicks. This depression then contributes to abnormal behaviour eg. partner aggression, feather plucking, failure to eat and as a result the bird then becomes exposed to all sorts of secondary problems. Any birds exhibiting such behaviour should be left to raise their own chicks in future. Another possibility arises when successive clutches are pulled at around the same age each time. Will the parents eventually automatically cease to feed around that point in time, rendering them useless for full-term breeding? If pulling is planned to be ongoing with a pair, then it is suggested that every second or third clutch, at least one chick be left until fledging.

Concerning the long-term consequences of successive generations never having the opportunity to raise their own chicks. Does the parent-chick interaction and stimulus in the early stages contribute to that chick's future parenting skills in a subtle psychological way? If so, and the handraised chick matures to produce chicks that again are handraised from the egg and so on, will the result be birds in several generations time that have no idea how to raise a chick? What about the mate-selection/acceptance of parent-reared chicks versus handraised chicks? Recent comparison studies suggest that indeed the natural pairing-up process is compromised by handrearing.

Aviculture is still very much in its infancy concerning evidence on the above, however many breeding facilities are now devoting considerable resources towards research and long-term monitoring of these behavioural trends. Although somewhat tenuous at this stage, early indications suggest that yes, the mentioned scenarios are more than possible. With this is mind, indiscriminate and perpetual removal of eggs and chicks is discouraged.

TROUBLESHOOTING

Handraising can be extremely satisfying and simple while everything is going smoothly and very frustrating when there is a problem in the nursery. Successful handraising is really all about prevention rather than cure, about maintaining a high level of hygiene, keeping accurate records, having the right equipment and feeding the right diet. It is not about waiting until things do go wrong before thinking, 'Why didn't I do something about that earlier?'.

In a nutshell, problems in the chick will manifest themselves in three ways:

- Poor weight gains or weight loss
- Slow crop movement
- Poor feeding response

A chick showing any or all of these symptoms should be brooded on its own, fed last and with a separate instrument until the cause is identified.

If a problem does arise, inevitably it will be despite your best efforts. It is then about being alert and catching the problem early when it is much easier to turn a chick around and initiate a rapid recovery. This alertness on the part of the feeder along with the quality of veterinary expertise available

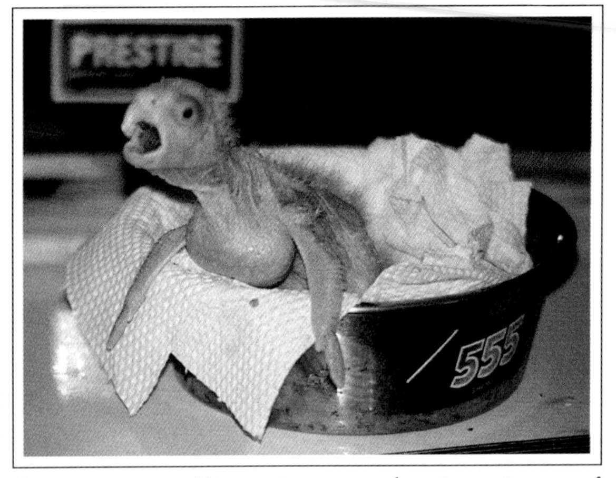

Incessant calling is a classic sign of underfeeding, at any age. However, this Western Long-billed Corella and other members of the Corella family exhibit this type of behaviour even when well fed.

today, will result in very few conditions that cannot be corrected. Following are the more commonly encountered conditions within the nursery. Note that most stem from poor management along with inexperience. As each problem is sourced and the necessary changes made, you step closer and closer towards uneventful handraising.

Poor Weight Gains

Following is a quick reference summary of the possible causes of poor weight gains or loss and attendant crop slowing.

- Underfeeding
- Incorrect diet for species concerned
- Incorrect brooder temperature
- Incorrect feed temperature
- Environmental stress
- Infections - bacterial, fungal or viral
- Aspiration
- Souring crop

Underfeeding

Underfeeding is undoubtedly the number one problem in the handraised chick, hence the emphasis on crop stretching and volume control thus far. Avoid underfeeding by ensuring that the total volume fed per 24 hours never goes backward and is consistently rising. If spoon feeding, ensure that the crop is nicely rounded and full each feed. As the chick grows so will its crop capacity.

Waiting for the crop to empty each time before re-feeding constitutes underfeeding. To keep the food moving and the chick growing at a rapid rate, feed on top of the remains of the previous feed. If however, the amount found left in the crop next feed begins creeping up past 25% then there is a problem developing, the mix was thickened too quickly or the volume increases were too much. Note this in your records and if things worsen it is possible to go back and see what changes could have been responsible for the situation.

A classic symptom of underfeeding is that the chick, like any baby that needs more food, will cry a lot. Small amounts in the crop do not leave a growing chick with that 'full' feeling, so keep the volumes rising. The well fed chick should do two things, eat and sleep. One species is the exception here. Chicks of the Corella family are extremely vocal and active, despite being well fed and in excellent health.

Another possible indication of underfeeding is where you have a chick that continually locks its beak onto the lip of the brooder container and pumps. This behaviour may or may not be a symptom of underfeeding. Obviously a chick that is not being fed enough will be continually searching for more food and will lock itself onto anything it thinks may produce food. However, this behaviour has also been observed in well fed, healthy and rapidly developing chicks. After addressing the possibility of underfeeding, change the brooder container for one where the rim is out of reach.

Be on the lookout for this behaviour occurring in your absence. Some chicks pump so vigorously that the brooder container literally hops around the brooder. Should you turn up for a feed to find the container in a new position then it is likely that this is occurring. Also check the chick regularly for the physical signs of such behaviour, a misaligned or out-of-shape beak. This frequent pumping on a hard rim will deform the beak of any chick but particularly in young ones where the beak is still quite soft. Depending on the extent of this behaviour it may also cause poor weight gains.

Slow Crop

Almost all problems, including underfeeding, will produce a slow moving crop to one degree or another. The longer formula remains in the crop the more it begins to sour, the more it sours the slower it moves and so the vicious circle goes. Hence, the minute a crop begins to slow there is the need to eliminate possible causes.

The first indication of a slowing crop will be when you find more than normal left in the crop. If volume records were kept then the possibility that the chick was simply overfed can be eliminated. Do not panic the first feed that this happens. Occasionally, for no apparent reason a particular crop full will be slow to empty, thereafter resuming normal emptying time. Feed a thinner mix and slightly less than normal, massage the crop around after the feed to loosen up the old food and stimulate the crop. Hopefully, next feed will see the crop almost empty again and indicate the slow crop was simply a hiccup in proceedings.

Many slow crops, where there is no detectable cause, respond well to Nilstat™ mixed in with the formula for two to three feeds at approximately 1ml per 50ml of formula. Also an excellent treatment for *Candida*, a yeast or fungus that will grow readily in the crop when food is not moving, Nilstat™ does appear to have a positive effect on crop movement. This, along with a good pinch of Probotic™, often puts crop motility back to where it should be. Try this as an initial treatment, however, if there is no improvement in the same day, look elsewhere for answers. Of course, if *Candida* is the problem, Nilstat™ will need to be continued.

There are other causes of slow crop movement. Feeding food too hot or too cold, especially with young chicks, will immediately slow digestion. It is recommended that a thermometer or probe be used to measure temperature of formula. Brooder temperatures will also directly affect crop motility and often a minor adjustment in this area will increase digestion rate. Black Cockatoos generally require lower brooder temperature than many other species and this temperature area may be a good place to start with slow crop in these birds.

Stress of any form needs to be considered. We already know that stress directly slows the crop, based on observation of chicks pulled from a nest environment and placed in the nursery. Small chicks brooded with larger chicks can often be knocked about and stressed, while chicks being brooded where there is a lot of activity may need moving to a quieter area. Should the brooder be one where heat is supplied from a light bulb that switches on and off, it will be necessary to diffuse the light in some way, as generally the darker the brooding environment the better. Another cause of crop slowing will be onset of infection and this is dealt with in detail shortly.

Crop Stasis

Crop stasis refers to a crop that has virtually stopped moving. A chick very rarely has a crop that shuts down instantly, except where physical trauma is involved. It is usually the final result of a crop progressively slowing down and the 'souring syndrome' kicking in. Total crop stasis is crisis time for the chick and manual crop emptying is essential before the chick dehydrates and dies. There are two ways of emptying the crop. Either manually express the contents or draw out the crop contents using the crop tube.

Manual expression involves holding the chick on its back and with its head lower than the crop, using the fingers to work the contents up the oesophageal canal and out the mouth. Not only is this extremely stressful to the chick, there is also the very real risk of aspiration and physical injury. Use this method as a last resort. Tubing, while still stressful to the chick is much less so and again becomes a lifesaver in this situation.

Until experienced, crop emptying by tube takes two people. One person holds the chick in the feeding position, while the other lowers the tube into the crop before drawing the contents into the syringe. Use one hand to hold the tip of the tube in the centre of the crop so as to avoid sucking the crop wall. Do not expect to remove the entire contents first time. Draw up as much as possible, then withdraw and expel the sour food into a bowl. Have on hand 1/2 cup of warm water with a pinch of baking soda (bicarbonate of soda) mixed in. Any sour crop turns acidic and the addition of baking soda will turn the crop more alkaline, improving motility. Half fill the crop with this water then while the tube is still down the chick, give the crop a quick massage before again sucking up as much as possible. Repeat this procedure one or two more times. By this stage the crop will be relatively empty and clean. This may sound like a lot of stress and work, however with a little experience the whole operation should take approximately two minutes.

Providing the cause of the stopped crop has been dealt with, then the crop should slowly resume its motility. Begin by feeding very thin formula with Spark™ and a good pinch of Probotic™, along with extra apple sauce. As the emptying time speeds up, begin thickening the formula. At each feed assess the remaining volume before re-feeding and if there is more than 25-30% of the prior feed still in the crop of the sick chick it pays to remove most of this with the tube first, to prevent the next crop full souring that much quicker.

Add a pinch of bicarbonate of soda to the fluid when crop rinsing. This changes the Ph level in the soured crop, turning it more alkaline.

Fungal Infection

Of all the infections it is possible for the growing chick to succumb to, *Candida albicans* is by far the most commonly encountered in the nursery. It is a fungal organism present in small numbers in the natural flora systems of most living creatures including humans, which under certain conditions can flourish. Some species appear to be more susceptible eg. White-tailed Black Cockatoos, however any compromised chick will be vulnerable and *Candida* is often a secondary problem arising from some other initial cause eg. stress, underfeeding or slow crop. It has been suggested that tube feeding causes a higher incidence of *Candida* infection, however many tube feeders would disagree and performed properly there would appear to be no correlation between tubing and this illness.

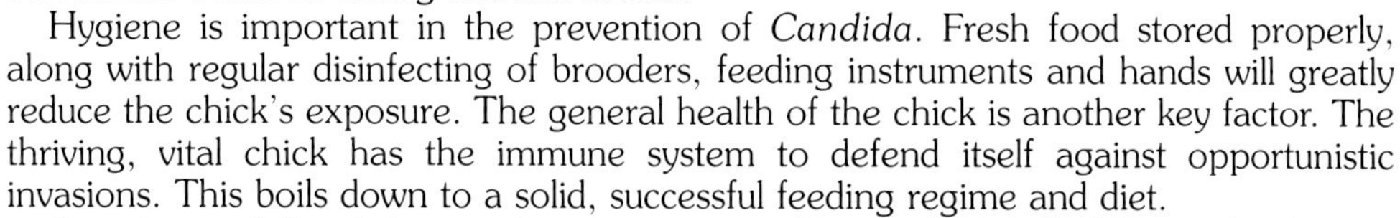

Hygiene is important in the prevention of *Candida*. Fresh food stored properly, along with regular disinfecting of brooders, feeding instruments and hands will greatly reduce the chick's exposure. The general health of the chick is another key factor. The thriving, vital chick has the immune system to defend itself against opportunistic invasions. This boils down to a solid, successful feeding regime and diet.

Symptoms of *Candida* are white spots or patches inside the chick's mouth. If this is the only site of infection it can be treated topically with Nilstat™ or other anti-fungal drugs as recommended by an avian veterinarian. Often however, *Candida* will be present either in the crop or further down the digestive tract without being visible in the mouth area. *Candida* in the crop will present itself as off-white areas visible through the crop wall and in more advanced cases will line much of the crop. In such cases a gentle rolling of the crop wall between the fingers will feel cheesy and thick.

An avian veterinarian can confirm the presence of *Candida* by taking swabs of faeces and crop contents and examining them under a microscope, and recommend the appropriate treatment.

Weaning time is a particularly susceptible period, especially with the larger species where weaning is prolonged. The stress of this period along with the weight loss weakens the chick, leaving it vulnerable. Incessant begging and heavy regurgitation are two signals of a chick with *Candida*. Another vulnerable time is when any chick or adult is on a course of antibiotics, as these drugs destroy many of the friendly bacteria in the system, reducing the bird's immunity. For this reason anti-fungal treatment should always run concurrently with antibiotic therapy and for a week or so afterwards.

Bacterial Infection

The risk of bacterial infection is largely controlled by the breeder and tight hygiene assists in preventing bacterial overgrowth. Generally, bacterial infection accounts for only a small percentage of nursery problems despite being one of the first areas many consider. There are many strains of bacteria, therefore indiscriminate use of antibiotics may do more harm than good. Consult an avian veterinarian for diagnosis and treatment.

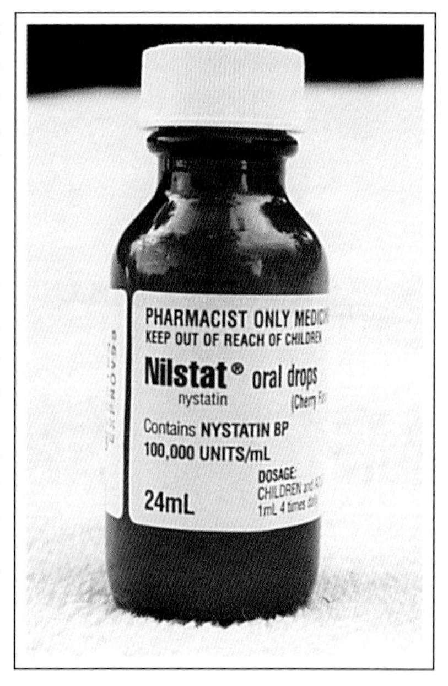

NilstatTM, an excellent anti-fungal medication used in the treatment of Candida.

Aspiration

Aspiration refers to the inhalation of fluid or formula into the airways. Slight aspiration will see a chick cough and sneeze for an hour or two before subsiding, however severe inhalation will result in difficult, rasping and heavy breathing. There is little that can be done except to keep the chick warm and hope that it recovers. In a worst case scenario the chick will inhale such a volume that death is almost immediate.

Aspiration is slightly more likely in the first few days of life where there is little or no feed response. Formula and fluids must be fed slowly and carefully until the chick does develop a response. Small chicks have aspirated by sleeping or leaning on a full crop of fluid, forcing it back up the neck. Pack tissues around the chick to help it sleep in an upright position.

Force feeding by spoon and expelling formula too fast by syringe are common causes of aspiration. Chicks need to breath periodically during feeding and if there is an excess of formula in the mouth, naturally some will be inhaled. The same applies to tubing, if the formula is expelled too fast or the tube is not in the crop far enough. The instant formula is observed flowing up the neck, withdraw the tube. Should the effects of aspiration continue for more than a couple of hours an avian veterinarian should be consulted concerning antibiotic treatment. The administering of oxygen and nebulizer treatment may also be undertaken by the veterinarian.

Regurgitation

Regurgitation of formula should never be ignored. If it is only slight and during peak/weaning period, there is little cause for concern and simple reduction in feed volume and/or frequency will often correct this. However, severe regurgitation during weaning, or regurgitation that was preceded by a slowing crop and low weight gains indicates that something a little more sinister is afoot.

The common cause is a crop infection, be it bacterial, fungal or both. Naturally a chick with an upset crop will want to reject any food placed in it. Once an infection is proven it is a case of administering the appropriate medication, thinning the mix a little to provide more fluids and giving the chick a few days to begin responding.

The simple cause of regurgitation is overfilling of the crop and any chick, even one a few days old, will want to get rid of some formula if fed too much. Formula fed too hot or too cold may also trigger vomiting and this is avoided by using a thermometer or temperature probe to accurately control feed temperature. Continual tubing may eventually irritate the oesophagus and crop if done incorrectly or if the flexible tube is not replaced when it hardens. Also, check the brooder temperature, because occasionally a chick brooded too hot will regurgitate.

Overstretched Crop

Overstretching, or pendulous crop as it is often referred to, is the result of increasing the volume too rapidly or by too much at once. The result is that the lower portion of the crop hangs over the top of the breast bone and the formula in that part of the crop will begin souring as it is unable to empty into the digestive tract. If the pendulous crop is not severe, remain on a volume setting for four to five days in the hope that the elasticity of the crop will return. During this period massage the crop well after each feed to keep the entire crop contents mixed and moving.

If the sagging is severe a crop support of some sort or *crop bra* will be necessary to elevate the lower crop above the keel bone. Broad band elastic can be purchased from a drapery shop and tailored to fit the chick. Stockings or lycra also make an excellent crop support. Tailor the material to fit the chick, putting in holes for the wings to help keep it in position.

Swallowed Objects

It is surprising what can end up in the crop of the handraised chick, particularly around weaning time and depending on what is swallowed, immediate removal will be necessary. Feed tubes that have come off the syringe need to be removed as soon as possible. Left too long it is possible for the tube to move down into the proventriculus making removal impossible without surgery.

Extended forceps used by the avian veterinarian to remove foreign objects from the crop.

Removal of any foreign object from the crop is not that difficult, however it will require two people. While one person holds the chick with its neck straight the other can manoeuvre whatever it is out of the crop and up the oesophagus. As soon as the item is visible at the rear of the mouth a blunt pair of tweezers can be used to reach in and remove it. If the crop was at least half full it is often easier to work with the object if the crop is emptied by tube first.

Woodchips when used as bedding occasionally end up in the crop, particularly with older chicks that are beginning to pick at anything and everything. Small chips will generally pass through, however larger splinters and chips will not. Such chips can be difficult to manoeuvre up the neck, particularly if pointed and a trip to the veterinarian becomes necessary.

Large pieces of corn cob, fruit and vegetables often break up and pass through over a couple of days. In this situation, the addition of digestive enzymes may help the process.

Splayed Legs

Chicks brooded on a smooth surface may find their legs continually slipping to the side as they try to sit or stand. If this is left uncorrected for too long the chick will develop splayed legs, a condition where the legs grow out on an angle rather than directly under the hip joints. Using correct bedding in the brooding environment will largely avoid this problem.

Poor nutrition has also been held responsible for leg splaying in some chicks, however commercial formulas have largely overcome this cause. Typical treatment is to tape the legs together with Co-flex™ or similar self-adhesive material in what would be the normal position. Also, add a calcium supplement to the diet such as Calcium

Sandoz™ or Calcivet™. It will also help to brood the chick in a relatively small container with padding around the legs to prevent side-slipping and in minor cases this together with calcium supplementation have corrected the problem without taping. This is a condition which may be corrected quite rapidly in the growing chick and a week or so of corrective treatment should see the legs back to normal. Rickets or deformed legs is a problem resulting from either a diet deficient in calcium or an absorption problem in the chick. Again, taping and calcium supplementation will sometimes correct this.

CONCLUSION

Incubation and handraising should be enjoyable experiences, therefore, do whatever it takes to make them just that. If you have a busy lifestyle, you need not miss out. Spend a little more money on an automatic turning incubator and learn how to crop tube. Avoid frustrating yourself with expensive losses and the hassles of a long-term sick chick, when a trip to the avian veterinarian will see the chick on the fast track to recovery. Do not pretend to be a veterinarian when you are not. Keep records and share with other breeders. If something is not working then ask advice, read books, search for answers and remember that it is not just you, everybody has their share of problems. Both the industry and your own approach are very much evolving situations. Experience and determination will reap the rewards.

BIBLIOGRAPHY

Parrot Incubation Procedures by Rick Jordan.

Parrots - Handfeeding & Nursery Management by Rick Jordan and Howard Voren.

Psittacine Aviculture by Richard M. Schubot, Kevin J. Clubb and Susan L. Clubb DVM.

Parrot Breeding Register by Rosemary Low.

Australian Parrots by Joseph M. Forshaw and William T. Cooper.

Tables

RELATIVE HUMIDITY (PERCENTAGE) TABLE															
Dry Bulb Temp°F	Difference between Dry and Wet Bulb Thermometer Readings °F														
	2	4	6	8	10	12	14	16	18	20	22	24	26	28	30
80	91	83	75	68	61	54	47	41	35	29	23	18	12	7	3
82	92	84	76	69	61	55	48	42	36	30	25	20	14	10	5
84	92	84	76	69	62	56	49	43	37	32	26	21	16	12	7
86	92	84	77	70	63	57	50	44	39	33	28	23	18	14	9
88	92	85	77	70	64	57	51	46	40	35	30	25	20	15	11
90	92	85	78	71	65	58	53	47	41	36	31	26	22	17	13
92	92	85	78	72	65	59	53	48	42	37	32	28	23	19	15
94	93	85	79	72	66	60	54	49	43	38	33	29	24	20	16
96	93	86	79	73	66	61	55	50	44	39	35	30	26	23	18
98	93	86	79	73	67	61	55	50	45	40	36	32	27	23	19
100	93	86	80	73	68	62	56	51	46	41	37	33	28	24	21
102	93	86	80	74	68	63	57	52	47	42	38	34	30	26	22
104	93	87	80	74	69	63	58	53	48	43	39	35	31	27	23
106	93	87	81	75	69	64	58	53	49	44	40	36	32	28	24
108	93	87	81	75	70	64	59	54	49	45	41	37	33	29	25
110	93	87	81	75	70	65	60	55	50	46	42	38	34	30	26

To calculate the relative humidity in the incubator or hatcher, subtract the wet bulb temperature from the dry bulb °F temperature in your incubator or hatcher. Locate this number on the top row of the chart, then come down the left column until you are on the line that corresponds to the dry bulb °F temperature of your unit. This will give you the approximate relative humidity to within one or two percent.

CONVERSION OF ° Celsius to ° Fahrenheit																			
°C	43.3	41.1	41.0	40.5	38.8	37.7	37.3	37.2	37.1	37.0	36.9	36.6	35.0	34.0	31.6	31.0	30.0	28.0	26.5
°F	110.0	106.0	105.8	105.0	101.8	100.0	99.1	99.0	98.8	98.6	98.5	98.0	95.0	93.2	89.0	87.8	86.0	82.4	79.7

Conversion Formula

Temp °F = (1.8 x °C) +32 • Temp °C = 0.555 x (°F-32)

WEIGHT GAIN TABLES

Following are weight gain tables of many of the more commonly handraised species. They are all records of chicks handraised from the egg. This is important to note because if your chick was parent fed initially for a period of time, then that chick's weights will probably be well ahead of those shown, having received the parental advantage. While these weight gains are typical for the species, should your weights be slightly behind there is no need to be concerned, the chick will still be on track to end up a fine specimen. Several factors that account for the difference in weight gains from breeder to breeder:

- Genetics
- Diet fed
- Volumes fed
- Experience of the handfeeder

If, however your chick's gains are significantly behind these weights then it DOES suggest that either the chick is being underfed, is on the wrong diet or is harbouring some sort of infection. Possible causes need to be addressed. Use these charts as a guide not as a definitive, to assess progress.

WEIGHT GAINS IN GRAMS

Days	Cockatiel	Scarlet-chested Parrot	Sun Conure	Red-browed Fig Parrot	Eclectus Parrot *E.r.polychloros*	Sulphur-crested White Cockatoo	Major Mitchell's Cockatoo	Red-tailed Black Cockatoo	Yellow-tailed Black Cockatoo	White-tailed Black Cockatoo	Glossy Black Cockatoo	Blue & Gold Macaw
1	3	3.1	6	2.8	16	18	11	24	29	18	15.3	20.6
2	3	3	6	3.2	16	21.5	12	26	30	18	16.1	21
3	4	2.9	6.5	3.6	18	25	15	32	34	20	15.6	22.8
4	5	2.8	7	4.6	23	30	20	34	38	24	15.7	20.7
5	6	3.1	7.3	5.1	32	34.5	24	35	42	24	16.5	23.3
6	8	3.4	8	6.4	41	42	30	39	48	26	17.5	24.6
7	10	3.8	9	7.9	49	53	34	46	60	32	18.9	30.6
8	11	4.2	10	8.6	58	68	41	54	67	38	21.1	35.1
9	13	4.6	10.5	10.6	69	80	51	62	77	44	23.7	38.2
10	15	5	13	11.4	75	97	62	71	86	52	25.7	39.5
11	18	5.8	14.5	12.8	90	115	70	83	95	60	28.7	47
12	21	6.5	16.5	14.9	96	137	75	94	99	70	31.7	54.4
13	23	7.6	18	16.4	108	160	80	105	111	84	34.6	60
14	26	8.1	20	20.4	122	186	85	118	126	96	40.2	67.8
15	29	9	23	20.8	138	215	101	132	150	108	42.1	78.8
16	33	10.5	26	21.8	150	248	118	144	160	118	44.3	88
17	38	11.5	29	24.7	171	280	129	160	177	136	52.2	102
18	43	13	33	26.3	189	314	143	172	182	156	55.6	110
19	48	14.2	37	28.2	200	348	147	181	191	166	55	120
20	58	15.6	41	29.5	220	385	158	192	201	180	56.3	128

WEIGHT GAINS IN GRAMS

Days	Cockatiel	Scarlet-chested Parrot	Sun Conure	Red-browed Fig Parrot	Eclectus Parrot *E.r.polychloros*	Sulphur-crested White Cockatoo	Major Mitchell's Cockatoo	Red-tailed Black Cockatoo	Yellow-tailed Black Cockatoo	White-tailed Black Cockatoo	Glossy Black Cockatoo	Blue & Gold Macaw
21	63	16.6	46	31.4	229	416	170	205	230	196	57.5	144
22	68	19.2	50	32.6	250	444	172	217	247	216	66	150
23	74	21	56	32.4	268	468	192	230	272	236	73.6	182
24	79	21.6	61	34.5	279	512	212	245	282	250	80.4	188
25	86	22.7	65	34.5	283	532	231	258	304	274	85	206
26	91	24.2	70	35.4	290	562	254	273	312	296	88	220
27	95	27	76	35.4	301	586	260	289	335	308	97	238
28	97	27.3	82	34.4	315	606	272	301	356	326	104.7	260
29	96	29.5	86	36.3 Peak	319	620	281	323	375	340	114	300
30	Peak	31	91	35.8	331	640	295	340	379	352	123.4	316
31		33.1	95	34.7	341	666	299	362	390	372	129.5	348
32		33	98	35.2	344	674	308	381	400	376	138.2	368
33		33.9 Peak	101	34.9	350	681	314	398	420	390	149.6	420
34		32.5	106	33.7	359	690	322	419	439	400	161.4	460
35		32.2	108	33.1	373	708	334	433	457	414	171.2	494
36		32.5	110	32.4	380	724	334	452	469	424	179.6	510
37		30.9	113	32.3	384	730	354	469	487	444	190.2	556
38		28.8	115	33.6	386	741	362	484	503	464	191.9	586
39		27.7	117	32	392	749	364	504	525	478	193.7	638
40		28.7	Peak	31.8	394	756	Peak	522	530	488	201.5	670

WEIGHT GAINS IN GRAMS

Days	Cockatiel	Scarlet-chested Parrot	Sun Conure	Red-browed Fig Parrot	Eclectus Parrot *E.r.polychloros*	Sulphur-crested White Cockatoo	Major Mitchell's Cockatoo	Red-tailed Black Cockatoo	Yellow-tailed Black Cockatoo	White-tailed Black Cockatoo	Glossy Black Cockatoo	Blue & Gold Macaw
41		27.5		31.3	Peak	762		536	560	492	203.6	688
42		29.2		30.1		769		548	564	496	212.8	692
43		29.3				775		567	588	500	221.7	758
44		30.1				780 Peak		581	592	518	229	764
45		32.8				779		588	599	528	235.8	784
46		Weaned				779		605	610	554	240	836
47								613	630	560	249.3	850
48								622	652	574	256.3	862
49								635	663	582	265.8	900
50								643	700	598	270.4	924
51								652	720	606	282.3	946
52								666	732	614	284	952
53								678	759	626	286.5	982
54								686	760	648	297.9	994
55								Peak	775	654	300	1012
56									790	674	305.9	1026
57									810	680	311.6	1068
58									825	688	310.5	1038
59									830	690	312.3	1038
60									832	692	315.5	1072

WEIGHT GAINS IN GRAMS

Days	Cockatiel	Scarlet-chested Parrot	Sun Conure	Red-browed Fig Parrot	Eclectus Parrot *E.r.polychloros*	Sulphur-crested White Cockatoo	Major Mitchell's Cockatoo	Red-tailed Black Cockatoo	Yellow-tailed Black Cockatoo	White-tailed Black Cockatoo	Glossy Black Cockatoo	Blue & Gold Macaw
61									845	694	317.8	1108
62									865	Peak	327.4	1074
63									870 Peak		329.7	1084
64									865		338.2	1100
												Peak
65											338.6	
66											332.7	
67											345.4	
68											347.2	
69											349	
70											344.5	
71											354.1	
72											362.1	
											Peak	
73											353.4	
74											350	
75											352.7	
76											347.3	

EXAMPLES OF CLUTCH SIZE, INCUBATION PERIOD, FLEDGING AGE AND ADULT WEIGHTS OF VARIOUS PARROTS

SPECIES	CLUTCH SIZE	INCUBATION PERIOD (days)	FLEDGING AGE (days)	ADULT WT. (grams)
Blue and Gold Macaw	2-4	24-26	90-100	1000-1200
Scarlet Macaw	2-4	25-26	98-105	900-1080
Green-winged Macaw	2-4	26-27	98-105	1200 (1040-1580)
Yellow-tailed Black Cockatoo	1-2	28-30	85-90	750-900
Red-tailed Black Cockatoo	1	28-30	75-95	600-850
White-tailed Black Cockatoo	1-2	28-30	75-95	600-700
Glossy Black Cockatoo	1-2	28-29	75-90	425-430
Palm Cockatoo	1	31-35	78-81	Cock 1000 Hen 800
Galah	3-5	23-25	45-55	330
Short-billed Corella	2-4	25-26	40-56	450-600
Western Long-billed Corella	3-5	23-25	45-56	550-700
Major Mitchell's Cockatoo	2-4	24-28	49-58	380-400
Gang Gang Cockatoo	2-3	28-30	50-60	280-300
Sulphur-crested White Cockatoo	2-3	25-28	62-84	850-900
Cockatiel	3-8	19-21	30-40	90
Eclectus - *E.r. polychloros*	2	28-30	72-80	380-410
Eclectus - *E.r. macgillivray*	2	28-30	72-80	550-600
Double Yellow-headed Amazon	2-4	26	60-65	500-530
African Grey Parrot	2-4	28-30	77-84	390-500
King Parrot	3-6	18-20	45-56	210-230
Crimson-winged Parrot	3-6	19-21	40-45	120-150
Superb (Barraband) Parrot	4-6	20-22	35-42	145-155
Regent Parrot	3-6	20-21	35-40	150
Princess Parrot	4-6	20-21	35-45	100-120
Red-capped Parrot	4-7	20-21	30-40	130
Mallee (Cloncurry) and Ringnecked Parrot	4-6	20-21	30-35	120-140
Port Lincoln and Twenty-eight Parrot	4-6	19-20	35-40	100-150
Eastern Rosella	4-9	19-22	30-35	100-120
Western Rosella	3-7	19-20	30-35	60-80
Northern Rosella	2-4	19-20	40-49	92-112
Tasmanian Rosella	4-5	19-22	30-35	140-165
Pale-headed Rosella	3-5	19-22	30-35	130
Crimson Rosella	4-7	19-22	30-35	123-169

EXAMPLES OF CLUTCH SIZE, INCUBATION PERIOD, FLEDGING AGE AND ADULT WEIGHTS OF VARIOUS PARROTS (continued)

SPECIES	CLUTCH SIZE	INCUBATION PERIOD (days)	FLEDGING AGE (days)	ADULT WT. (grams)
Adelaide Rosella	4-7	19-22	30-35	112-165
Yellow Rosella	4-5	19-22	30-35	110-135
Red-rumped Parrot	4-7	19-20	30-35	65-70
Mulga Parrot	4-7	19-20	28-35	56-65
Blue-bonnet Parrot (Red-vented)	4-9	19-20	30-40	90-100
Golden-shouldered Parrot	3-6	19-20	28-35	55-60
Hooded Parrot	3-6	19-20	30-40	55-60
Kakariki	5-12	19-20	35-40	95-100
Indian Ringnecked Parrot	4-5	21-24	45-50	116-140
Alexandrine Parrot	3-4	24	50-60	250-258
Plum-headed Parrot	4-5	23-24	45-50	90
Slaty-headed Parrot	4-5	23-24	45-50	100-120
Derbyan Parrot	2-3	23-24	50-55	270-300
Malabar Parrot	3-4	24-27	49-56	90
Moustache Parrot	3-4	21-24	45-50	130
Sun Conure	3-4	24-28	49-56	100-120
Janday Conure	3-6	22-25	49-56	100-120
Nanday Conure	3-5	21-23	49-56	110-130
Peach-fronted Conure	2-4	23-24	49-56	80
Bourke's Parrot	3-6	18-20	23-35	48-50
Rock Parrot	3-5	19-20	35-40	50-52
Elegant Parrot	4-5	18-21	30-40	45-50
Scarlet-chested Parrot	4-6	17-22	26-32	40
Turquoisine Parrot	3-5	19-21	24-34	40
Blue-winged Parrot	4-5	19-20	31-35	50-55
Budgerigar	4-7	18	31-35	35
Swift Parrot	3-5	20	42	70
Peachface Lovebird	3-7	18-24	42-56	55
Rainbow Lorikeet	2	24	55-60	130-150
Scaly-breasted Lorikeet	2	22	55-60	85
Varied Lorikeet	2-5	22	40	60
Musk Lorikeet	2	24	45-48	60
Purple-crowned Lorikeet	3-4	19-22	45-60	45-50
Little Lorikeet	3-5	20-22	40-45	45

The Acclaimed 'A Guide to...' series.

- **A Guide to Australian Grassfinches**
  The popularity of Australian Grassfinches worldwide is largely due to the hardiness of these tiny, gregarious and colourful birds. The 18 members of the Grassfinch family Estrildae recognised in Australia are featured in detail. Diagrams indicating visual differences and some 160 coloured photographs support the 80 pages of text. A must for every finch breeder's library.

  ISBN NUMBER 0 9587102 2 8 Author: Russell Kingston

- **A Guide to Neophema and Psephotus Grass Parrots and Their Mutations (Revised Edition)**
  This 88-page title features over 160 full colour images of mutations in the *Neophema* and *Psephotus* grass parrot group. There are examples of breeding expectations, housing, feeding and management.

  ISBN NUMBER 0 9587102 4 4 Author: Toby Martin

- **A Guide to Asiatic Parrots and Their Mutations (Revised Edition)**
  Containing over 70 colour images of Asiatic parrot mutations, this 88-page title also features genetic tables and information on nutrition, housing, breeding and mutations.

  ISBN NUMBER 0 9587102 5 2 Authors: Syd & Jack Smith

- **A Guide to Australian Long and Broad-tailed Parrots and New Zealand Kakarikis**
  This 88-page full colour title features beautiful photography throughout. Each of the 12 species is featured in its own chapter with distribution maps and general information specific to that species including management, diet and nutrition, housing requirements, breeding, handrearing, sexing and mutations.

  ISBN NUMBER 0 9587455 3 6 Author: ABK Publications

- **A Guide to Rosellas and Their Mutations**
  This full colour title features the general management, care and breeding of the *Platycercus* genus. Breeding expectations, including genetic tables and mutations, are discussed for each species and their subspecies. Beautiful photography throughout.

  ISBN NUMBER 0 9587455 5 2 Author: ABK Publications

- **A Guide to Gouldian Finches and Their Mutations (Revised Edition)**
  This 160-page complete revision features concise information on Gouldian Finches in the Wild and in Captivity, including Housing, Nutrition, Breeding, Mutation and Colour Breeding, Health and Disease. Supported by over 330 images, including an extensive selection of mutations. Authors have contributed to various chapters within the book.

  ISBN NUMBER 0 9750817 1 3 Author: ABK Publications

A Guide to Cockatiels and Their Mutations
Written by two of Australia's foremost Cockatiel breeders, this 96-page title features beautiful colour photography, including all known mutations. Excellent easy-to-read information covers the care, management, housing and breeding of these popular birds.

ISBN NUMBER 0 9587455 8 7
Authors: Peggy Cross and Diana Andersen

A Guide to Pigeons, Doves and Quail
A world first in aviculture, this 184-page title covers all species in this group available to the Australian aviculturist. Stunning colour photography throughout is supported by precise, easy-to-read information on the care, management, health and breeding of these unique birds.

ISBN NUMBER 0 6462305 8 1 Author: Dr Danny Brown

A Guide to Lories and Lorikeets (Revised Edition)
Completely reformatted and revised including new sections on Lories and Lorikeets as Pets, Diseases and Disorders and Colour Mutations and Breeding Expectations, this 152-page title is bigger, better and more colourful than the highly successful original edition. Peter's exceptional photography again is beautifully supportive of the informative text which together make this a must-have title.

ISBN NUMBER 0 9577024 4 2 Author: Peter Odekerken

A Guide to Basic Health and Disease in Birds (Revised Edition)
Since its first publication in 1996, this 112-page title has proven to be one of the most sought after and respected titles worldwide in this generic range of avian publications. It is a credit to the author, Dr Michael Cannon. His devotion and concern for all aspects of avian health and husbandry have again been reflected in this revised edition.

ISBN NUMBER 0 9577024 5 0 Author: Dr Michael Cannon

A Guide to Eclectus Parrots (Revised Edition)
This 160-page title features a comprehensive description of all 10 subspecies. Chapters include: Taxonomy and Identification, Eclectus in the Wild, Eclectus in Captivity—as Pet and Aviary Birds, Housing, Feeding, Breeding, Artificial Incubation and Handraising, Troubleshooting and Symptoms of Breeding Failure, Taming and Training, Colour Mutations and Genetics, Diseases and Disorders. Featuring over 250 colour photos, this 160-page title is available in both soft and hardcover format.

ISBN NUMBER 0 9750817 0 5
Authors: Dr Rob Marshall BVSc MACVSc (Avian Health) and Ian Ward

A Guide to Pet & Companion Birds
This informative and often amusing 'introduction to bird keeping' appeals not only to the novice or want-to-be bird keeper, but also to the seasoned aviculturist looking for a refresher on the basics. This 96-page full colour book also guides you through the growing pains of increasing your bird family, including what to do when your birds have gone forth and multiplied.

ISBN NUMBER 0 9587266 1 2
Authors: Ray Dorge and Gail Sibley

- **A Guide to Pheasants & Waterfowl**
Author of the highly regarded *A Guide to Pigeons, Doves & Quail*, Dr Danny Brown has produced this superlative title on pheasants and waterfowl. The informative, 248-page, easy-to-read text is lavishly supported with beautiful colour images throughout. Covering all aspects of caring, housing, management and breeding of these unique birds, this title is a credit to the author and an ideal reference source.

ISBN NUMBER 0 9587102 3 6 Author: Dr Danny Brown

- **A Guide to Australian White Cockatoos**
Richly illustrated and full of practical hints, this well-researched, well-written book features facets of the author's personal experience with the Australian White Cockatoo family which shine throughout its 112 full colour pages.

ISBN NUMBER 0 9577024 1 8 Author: Chris Hunt

- **A Guide to Zebra Finches**
This title features 96 pages of easy-to-read, highly informative text, including all currently recognised Australian colour varieties on these internationally popular endemic birds. Full colour throughout, this book is a must for any Zebra Finch enthusiast.

ISBN NUMBER 0 9577024 2 6
Authors: Milton, John and Joan Lewis

- **A Guide to Popular Conures**
Featuring 13 of the most popular conures kept as pet or aviary birds throughout the world, this 112-page title is packed with easy-to-read, highly usable information and features superb full colour photographs throughout. Combined with their own experiences, Ray Dorge and Gail Sibley have recorded the results of an extensive research of large conure breeders. This title is sure to satisfy all fanciers and breeders of these wonderful parrots.

ISBN NUMBER 0 9577024 3 4
Authors: Ray Dorge & Gail Sibley

- **A Guide to Macaws as Pet and Aviary Birds**
Recognised internationally as one of the world's most accomplished and talented aviculturists, published author and speaker, Rick Jordan has produced the perfect companion book for anybody interested in macaws, be it as a pet or as breeders. Featuring spectacular full colour photography throughout, this 136-page soft cover title is packed with valuable and highly useable information.

ISBN NUMBER 0 9577024 9 3 Author: Rick Jordan

- **A Guide to Colour Mutations and Genetics in Parrots**
This title, that has taken Australian author, Dr Terry Martin BVSc, some five years to complete, is the most definitive, collective work ever attempted on this intriguing and contentious subject. Drawing on information from specialist parrot mutation breeders from all over the world, Terry Martin has collated over 700 colour photographs in this 296-page soft and hard cover title within text that is both approachable and easily understood.

ISBN NUMBER 0 9577024 6 9 Author: Dr Terry Martin